Experimental low-temperature physics

MACMILLAN PHYSICAL SCIENCE

Series advisers

Physics titles: Dr R L Havill, *University of Sheffield*
Dr A K Walton, *University of Sheffield*

Chemistry titles: Dr D M Adams, *University of Leicester*
Dr M Green, *University of York*

Titles in the series

Group Theory for Chemists, *G Davidson*
Thermal Physics, *M Sprackling*
Lanthanides and Actinides, *S Cotton*
Introduction to Electrochemistry, *D B Hibbert*

MACMILLAN PHYSICAL SCIENCE SERIES

Experimental low-temperature physics

Anthony Kent

Dept of Physics, University of Nottingham

First published 1993 by
THE MACMILLAN PRESS LTD
Houndmills, Basingstoke, Hampshire RG21 2XS
and London

Library of Congress Cataloging-in-Publication Data
Kent, Tony.
Experimental low temperature physics / Tony Kent.
p. cm. — (Macmillan physical science series)
Includes bibliographical references and index.
ISBN 978-1-56396-030-7
1. Low temperatures. I. Title. II. Series.
QC278.K46 1993
536'.56—dc20 92–10419
CIP

Contents

Preface

Many university physics departments now offer a short series of lectures on low-temperature physics. Usually these are given as a final-year option, or as part of a course on thermal physics. Low-temperature physics pulls together strands from a number of essential core subjects: thermodynamics and statistical physics, quantum physics and solid-state physics. It also provides an interesting framework in which to illustrate and expand on the theoretical concepts taught in previous years. Applications of low temperatures are becoming widespread in research and industry, and a basic grounding in the subject is considered desirable for all serious students of science and technology.

This textbook fills the need for an uncomplicated introduction to experimental aspects of low-temperature physics, concentrating on the methods used to achieve low temperatures in the laboratory. I have avoided the detailed theoretical treatments found in other books, which are primarily aimed at researchers in the subject. Some basic experimental techniques are covered, but at a level that might be of use to an undergraduate project student or a first-year research student using, rather than researching into, low-temperature techniques. I have included theory only where it is essential to illustrate the point being made and without using mathematics or introducing concepts that the average third-year undergraduate student will not have met before.

In Chapters 2 and 3, I review some of the basic physical properties of matter at low temperatures. Detailed theoretical explanations of the properties are not given, because these may be found in other standard textbooks on thermal and solid-state physics. Instead, I pay particular attention to those properties that are relevant to the material of later chapters. In Chapters 4, 5 and 6 I describe the methods that are used to attain low temperatures in the laboratory, starting from around 300 K and ending at less than a microkelvin. Thermometry goes hand in hand with being able to attain low temperatures. In Chapter 7, I describe the methods, both primary and secondary, used to measure low temperatures. Finally, Chapter 8 contains a few 'hints and tips' for the would-be low-temperature experimenter.

I am grateful to the following for granting me permission to reproduce copyrighted figures: J. G. M. Armitage, University of St. Andrews (Figure 3.5(b)); Oxford University Press (Figures 3.10–3.12); McGraw-Hill (Figure 4.5); Oxford Instruments Ltd (Figures 7.8–7.10); Cambridge University Press (Figures 5.7, 5.8 and 7.15(c); Academic Press (Table 7.2); and Linde Cryogenics Ltd (Figure 4.2).

Every effort has been made to trace all copyright holders, but if any have been inadvertently overlooked, the publishers will be pleased to make the necessary arrangements at the first opportunity.

Nottingham, 1992 AJK

1 Introduction

1.1 The concept of temperature

What is a *low temperature*? Before we can start to consider the properties of matter and experimental techniques at low temperatures we must answer this question. Everyone has an opinion on the matter. Most would describe the inside of a domestic food freezer as being at a rather low temperature. However, an inhabitant of the polar regions would probably find it nothing out of the ordinary. The point is that there is really no absolute answer to this question and that it depends on the context in which we are speaking. Certainly, as far as this book is concerned, the inside of the domestic food freezer is at a very high temperature! This question also raises another: What exactly do we mean by the *temperature* of something?

Temperature is a statistical concept applying only to systems containing a large number of particles. It would be meaningless to describe a single isolated particle as having a temperature. In high-school physics the concept of temperature is often introduced by the following statement: 'When two bodies are brought into thermal contact, heat flows from the body at the higher temperature to the body at the lower temperature'. In other words, when the two bodies are in thermal equilibrium their temperatures are equal. This is a statement of the zeroth law of thermodynamics. The fact that it is true makes possible the measurement of temperature: If a thermometer is brought into contact with a body at a certain temperature, then given enough time the thermometer and the body will reach the same temperature. This could well be different from the initial temperature of the body, because in order to achieve equilibrium, heat must flow between it and the thermometer. Thermometers work by having some physical quantity that changes with temperature in an easily observable way, such as, for example, the length of a column of mercury or alcohol, which could be calibrated in any arbitrary manner. In order to standardise measurements made in different laboratories, certain scales of temperature have been constructed around fixed calibration points, e.g. the Celsius scale, which uses the ice point and the steam point of pure water at atmospheric pressure to define 0 °C and 100 °C, respectively.

A rather better definition of the temperature, and one which allows it to be regarded as a real physical quantity, is that it is a function of the thermal energy of the system, i.e. the energy associated with the random motions of the constituent particles in it. This is often called the thermodynamic temperature. It leads to the definition of an absolute scale of temperature so that the thermal energy of any system is zero at the absolute zero of temperature. In a gas the thermal energy is in the form of the random, motions of the molecules and is proportional to the mean kinetic energy. A gas thermometer works on the principle that the pressure a gas exerts on the walls of its container depends on the speed of the molecules striking them. Let us imagine that we have a constant-volume gas thermometer which contains a perfect (or ideal) gas, that is, one in which there are absolutely no intermolecular forces. In order to calibrate the thermometer we plot a graph of pressure against temperature between, say, 0 and 100 °C. A straight line will be obtained, which may be extrapolated back to zero pressure where the mean kinetic energy (and hence the thermal energy) of the gas must be zero and we will discover that this point corresponds to a temperature of minus 273.15 °C, the absolute zero of temperature. The Kelvin scale of temperature is constructed such that absolute zero is zero Kelvin (0 K), and the divisions are the same size as in the Celsius scale, hence on this scale the boiling point of water is +373.15 K. However, this still does not put us in a position where we can say with any physical meaning what a *low temperature* is. We can define it only as being a temperature that is lower than some arbitrarily fixed point such as 0 °C or 273.15 K.

We could turn to the quantum-mechanical description of matter to help resolve this problem. The theory of quantum mechanics is based on the fact that for any system there are a set of discrete physical states, quantum states, that it is allowed to be in. To each of these states corresponds a particular energy level of the system. The lowest energy the system can have, corresponding to its ground state, is not necessarily zero, but has some value which, like all of the energy levels, depends only on the precise microscopic nature of the system under consideration. At the absolute zero of temperature all systems will be in their lowest energy state because there is no extra energy in the form of thermal energy available to lift them out of it. We will not allow the fact that absolute zero is unattainable in reality (which follows from the third law of thermodynamics) to worry us at this stage. As the system is warmed from 0 K, the higher energy states become occupied. The probability of occupancy of the excited states relative to that of the ground state is proportional to the absolute temperature. Further increases in the temperature will lead to yet higher levels being occupied until eventually, at sufficiently high temperatures, the thermal energy of the system is very much greater than the ground state energy, as would be the case for a macroscopic volume of an ordinary gas at room temperature. When in a 'thought experiment' such as this the temperature is increased from absolute zero, it must arrive at a point at which the thermal energy is comparable with the ground state energy. At this point, in a macroscopic sample of many constituent particles, the thermal de Broglie wavelength of the particles becomes comparable to the average interparticle separation. Below this temperature the quantum-mechanical nature of the system is expected to be most evident. This might appear to be a suitable point to use in order to differentiate between high and low temperatures.

This definition of a low temperature has some problems which are best illustrated by a couple of examples:

1. Free conduction electrons in a metal at room temperature

The thermal de Broglie wavelength of a particle, mass m, is given by

$$\lambda_{dB} = h/mv.$$

Where h is Planck's constant (6.6262×10^{-34} J s), v is the root mean square thermal velocity at temperature T, $v = (3kT/m)^{1/2}$, and k is Boltzmann's constant (1.38062×10^{-23} J K^{-1}). In a typical metal there are about 10^{29} free conduction electrons per cubic metre. Each electron, therefore, occupies a volume of 10^{-29} m^3 which implies an average interparticle separation of about 2×10^{-10} m. Equating this to λ_{dB} gives $T \approx 2 \times 10^5$ K! It is clear that even room temperature at 300 K is a very low temperature in these terms.

2. Liquid nitrogen at 77 K

The density of liquid nitrogen at its normal boiling point of 77 K, is 810 kg m^{-3} and the mass of a nitrogen molecule is 4.65×10^{-26} kg. These figures imply an average interparticle separation of 3.9×10^{-10} m. The temperature which gives λ_{dB} equal to this is only 1.5 K and so liquid nitrogen at 77 K should be considered to be at a rather high temperature!

It appears that whether a particular temperature is regarded as being high or low in this context depends on the system under consideration.

So, what temperature do we call a low temperature? The answer, as far as the terms of reference of this book are concerned, is that the line between high and low temperatures will be drawn according to the technology required to achieve and work at those temperatures. Low temperatures will be those that require the use of liquefied gases with boiling points near to or below the boiling point of liquid nitrogen to be reached, or those temperatures lower than can be obtained by using conventional refrigeration techniques. This technology is called *cryogenics* from the Greek words *cryos* (meaning cold) and *genes* (meaning generated from).

Figure 1.1 shows the range of temperature currently attainable in the laboratory. Notice the wide range of temperature, which is conveniently presented on a logarithmic scale. Highlighted are certain reference points, including important landmarks in the development of low-temperature technology.

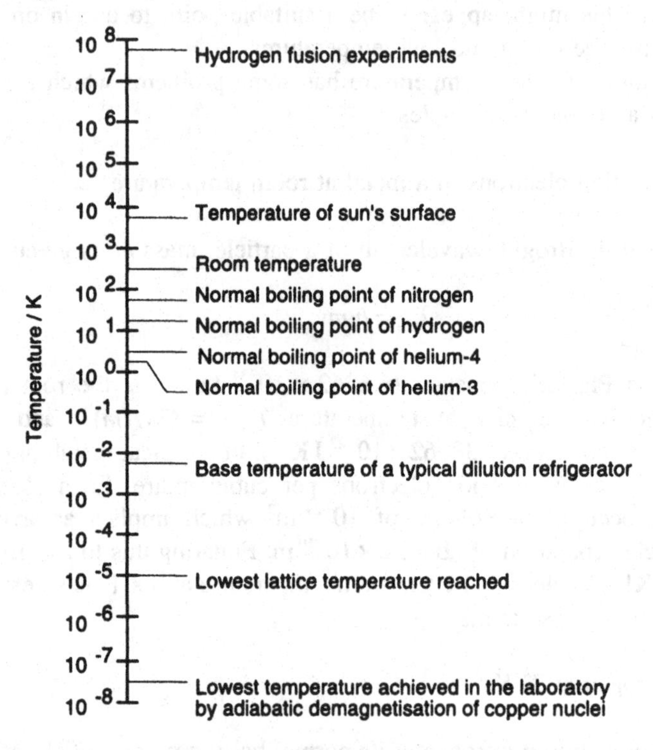

Figure 1.1 *Range of temperatures achievable in the laboratory. The vast range, 16 orders of magnitude, necessitates the use of a logarithmic scale*

1.2 The laws of thermodynamics

References have been made already to two of the laws of thermodynamics. Thermodynamics is the subject which deals with the concepts of temperature, entropy and heat, concepts which we will encounter frequently in the following chapters. It is outside the scope of this book to delve in great detail into thermodynamics, other than to say that it is a macroscopic description of matter based on a few fundamental statements or laws. For a fuller discussion the reader is referred to some of the excellent texts on the subject contained in the bibliography at the end of this chapter. However, in view of their importance in the subjects of later chapters, a short summary of the thermodynamic laws is given here.

1.2.1 The zeroth law

This is often stated as follows: If two systems are in thermal equilibrium with a third then they are in equilibrium with each other.

This rather trivial statement means that it is possible to measure temperature by means of thermometers as discussed in Section 1.1.

1.2.2 The first law

This law can be thought of as a statement of the conservation of energy in thermodynamic systems. A system in equilibrium can be said to possess an internal energy, E, if the system is taken from an initial macrostate to a final macrostate adiabatically, that is, it is not allowed to exchange heat with its surroundings, then the work done on the system, W, is the difference between the initial and final internal energies, $\Delta E = E_f - E_i$. If the system is also allowed to exchange heat Q with the surroundings, then according to the conservation of energy,

$$\Delta E - W - Q = 0.$$

This is the mathematical statement of the first law, which may be expressed in differential form in the situation when the initial and final states are brought very close together:

$$\Delta Q = + \mathrm{d}E - \Delta W$$

1.2.3 The second law

It is a fact (and a rather disconcerting one when it comes to attaining low temperatures) that it is impossible to transfer heat from a cold to a hot body without expending work. This is known as the Clausius statement of the second law of thermodynamics. It means that it is impossible to construct the ideal refrigerator; they all require an external source of power to operate. An alternative and equivalent statement of the second law, attributable to Kelvin, is that it is impossible to construct a perfect heat engine, i.e. one which can take heat from a reservoir and convert it entirely into work; the cooling towers of modern power stations are a tribute to this. The converse is possible however: any amount of work can be converted to heat. Both of these statements imply that in addition to its internal energy, E, a thermodynamic system is also characterised by a quantity known as the entropy, S, which must possess the following properties:

1 The entropy of a thermally isolated system undergoing a change of state tends to increase, i.e.

$$\Delta S \geq 0.$$

2 If, on the other hand, the system is not isolated and undergoes an infinitesimal and reversible change of state involving the exchange of heat, dQ, with the surroundings then the entropy change is defined as follows:

$$dS = dQ/T.$$

The entropy can be regarded as a measure of the degree of disorder in the system.

1.2.4 The third law

The entropy, S, of a system tends to a constant value, S_0, independent of all parameters of the system as the temperature approaches absolute zero. This statement has important implications for us; it means that it is impossible to achieve the absolute zero of temperature in any finite number of steps. We can demonstrate this by considering the use of a hypothetical perfect gas as a refrigerant: Figure 1.2 shows schematic T–S curves for the gas at two constant pressures, p_1 and p_2; in accordance with the third law, the two isobars converge to S_0 at zero temperature. Cooling may be achieved by an isothermal compression of the gas from p_1 to p_2, followed by a reversible adiabatic expansion back to p_1. During the adiabatic expansion, no heat is exchanged with the surroundings, $dQ = 0$, hence from the second law $dS = 0$, the gas performs work at the expense of its internal energy and cools isentropically from T_i to T_f. In principle, these operations, isothermal compression followed by adiabatic expansion, could be repeated continually in an attempt to reduce the final temperature to absolute zero. However, the convergence of the two curves means that as we approach absolute zero the size of the steps $T_f - T_i$ tends to zero, and so we would require an infinite number of them to actually reach the goal $T = 0$ K.

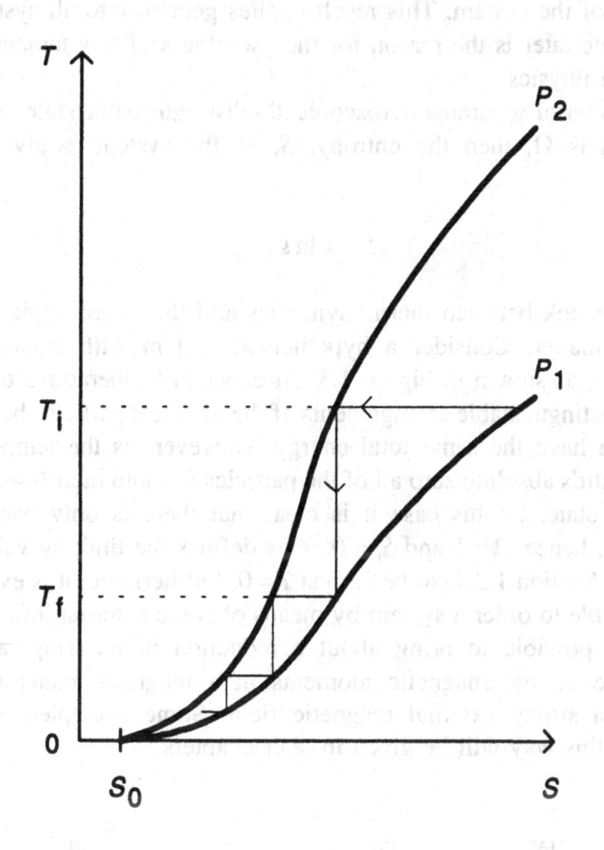

Figure 1.2 *Schematic T–S curves at two constant pressures for a hypothetical coolant. The curves converge at T = 0 in accordance with the third law of thermodynamics.*

1.3 Order and disorder and low temperatures

Consider a sample of nitrogen gas, initially at room temperature. If it is cooled down to 77 K it liquefies, and upon further cooling to 59 K it solidifies. Each phase change, gas to liquid and liquid to solid, involves an ordering of the system, the solid obviously being the more ordered state. If we look inside the solid (there are experimental techniques that allow us to do just this) at 59 K then we will observe a large amount of molecular motion, vibrations and thermally created interstitials and vacancies. Further cooling of the solid slows all of these motions which, if it was possible to actually achieve 0 K, would eventually cease altogether, apart from motions associated with the finite—zero—point energy of the system. So, cooling a system to low temperatures decreases its entropy; in other words, it leads to an ordering of the system. This result applies generally to all systems, and as we will see later is the reason for the usefulness of low temperatures in experimental physics.

If the number of separate microscopically distinguishable states accessible to a system is Ω, then the entropy, S, of the system is given by the expression

$$S = k \ln \Omega.$$

This forms a link between thermodynamics and the microscopic statistical theories of matter. Consider a hypothetical system with equally spaced energy levels, as shown in Figure 1.3. At elevated temperatures there are a number of distinguishable arrangements of the identical particles between the levels which have the same total energy. However, as the temperature is reduced towards absolute zero all of the particles fall into their lowest energy (or ground) state. In this case it is clear that there is only one possible arrangement, hence $\Omega = 1$ and $S_0 = 0$. This defines the limiting value of the entropy (see Section 1.2.4) to be zero at $T = 0$. Furthermore, it is evident that if we were able to order a system by means of some external influence then it might be possible to bring about a reduction in its temperature. For example, the atomic magnetic moments in a magnetic material can be aligned by a strong external magnetic field. Some examples of cooling achieved in this way will be given in later chapters.

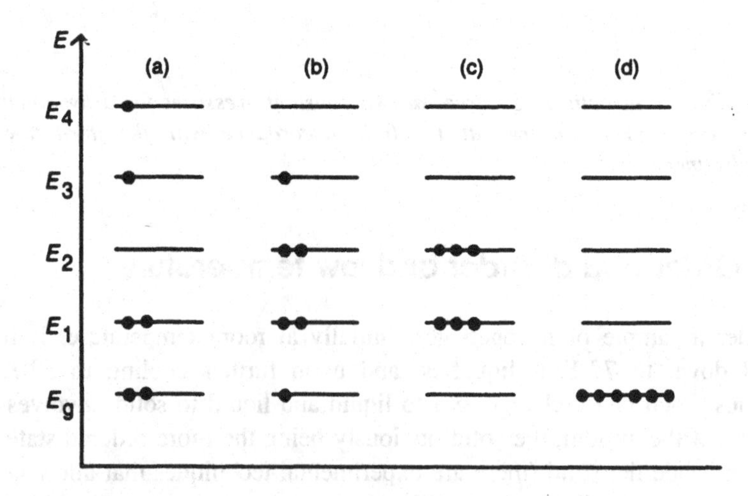

Figure 1.3 *System comprising six particles and equally spaced energy levels. States (a), (b) and (c) are all distinguishable but have the same energy. There is only one arrangement, (d), corresponding to the ground state*

1.4 A brief history of low-temperature physics

Although it is impossible to do justice to all the contributions to the development of this field, there are a few landmarks worthy of special mention. In 1877 Cailletet in France and Pictet in Switzerland (working independently) first liquefied oxygen. They were able to perform the first low-temperature experiments with a study of the properties of oxygen below its critical temperature, that is, the temperature at which the liquid and dense gas phases are no longer distinguishable and there exists just one fluid phase of the substance. A serious problem at the time, which threatened to hold up progress, was the storage of the cold liquids. This was solved in 1882 when Dewar developed the vacuum flask, a double-walled vessel with a vacuum between the walls to minimise heat transfer by thermal conduction and silvered on the inside to minimise heat transfer by radiation. Dewar was also the first to liquefy hydrogen (boiling point 20.4 K) by isenthalpic expansion, a process that makes the gas perform work against its own internal intermolecular forces, at the expense of its internal energy. The next major step was taken in 1908 when Kamerlingh Onnes was the first to liquefy helium. Working in Leiden, Onnes' group were the first to discover superfluidity in liquid helium and in 1913 superconductivity in mercury. Thirty to forty years of experiments followed on the low-temperature properties of matter, but progress was hindered by the fact that only a few laboratories had the equipment to liquefy helium, a process fraught with danger at the time, owing to the copious quantities of liquid hydrogen required.

The next major step was taken during the early 1940s when in the USA S. C. Collins designed a helium liquefier which operated without the need for liquid hydrogen as a pre-coolant prior to the isenthalpic or, as it is often called, the Joule-Thompson expansion stage. The system (to be described in greater detail in a later chapter) used expansion engines to pre-cool the helium gas. It formed the basis of small commercial liquefiers produced by A. D. Little Inc. and this predictably led to the rapid expansion of the field of low-temperature physics, as more laboratories and institutions could acquire the apparatus to liquefy helium.

Researchers striving for still lower temperatures developed techniques such as ^{3}He-^{4}He dilution refrigeration and magnetic cooling. This last technique, first demonstrated by Onnes, has been used to achieve temperatures as low as 5×10^{-8} K with copper nuclei, a figure which will almost certainly have been bettered by the time this book is read. It is now quite common, although it still involves much time and effort, to carry out experiments at temperatures down to a few millikelvins using commercially available equipment. None of this would have been possible but for the pioneering work of Onnes in producing and using liquid helium, which earned him a Nobel prize in Physics.

1.5 Applications of low temperatures

The main part of this book will be concerned with the techniques through which low temperatures are achieved. At this stage, it is worth while considering just a few of the applications in science and technology that low temperatures are put to. As discussed in Section 1.2, when temperatures are lowered, the thermal energy of the system eventually becomes comparable to the ground state energy and the quantum-mechanical nature of the system becomes most apparent. For this reason, low temperatures are widely used in the study of quantum phenomena in condensed matter, especially liquid helium and semiconductors. This forms a large part of the basic scientific research in academic and industrial laboratories throughout the world. Surprisingly, low-temperature studies of matter can be used in order to learn more about how materials behave at much higher temperatures. For example, semiconductor devices often work at temperatures above room temperature. However, by cooling them down to remove the crystal lattice vibrations, the effects of impurities etc. on the electronic properties can be studied independently. The vibrations can be put back in a controlled manner by warming the sample slightly, so that their effect can be studied in more detail.

The study and use of superconductive phenomena is a major application of low-temperature techniques. Much of the commercial effort is directed in this area. Superconducting magnets, in their simplest form a solenoid of superconducting wire cooled in liquid helium, are widely used in basic research. They enable high magnetic fields to be achieved and maintained with a low power consumption. Large magnets with a bore sufficient to accommodate a human body are used in medical imaging. Other applications of superconductors include superconductive quantum interference devices (SQUIDS), which find application as highly sensitive magnetometers, and Josephson junctions, which might find application in superconducting electronics. More recently, superconductors which operate at and above the temperature of liquid nitrogen have been discovered and research into applications for these is expanding rapidly.

Rather more straightforward applications of low temperatures include the storage and transport of gases such as nitrogen and helium. Rocket fuels are stored in their liquid state, because much larger quantities of the fuel can be stored at a lower total weight (fuel plus storage vessel) than can be stored as a compressed gas. Storage of biological materials such as blood is carried out under liquid nitrogen. In scientific research, liquefied gases are used to cool nuclear particle and optical detectors. Electronic amplifiers used in some critical applications, such as radio astronomy, are designed to be operated at cryogenic temperatures, where the noise generated in the electronic devices is lower.

Bibliography

Suggested background reading on thermodynamics and statistical physics

Landau, L. D. and Lifshitz, E. M. (1959). *Statistical Physics* (Reading, Mass.: Addison-Wesley)

Pippard, A. B. (1957). *The Elements of Classical Thermodynamics* (London: Cambridge University Press)

Reif, F. (1965). *The Fundamentals of Statistical and Thermal Physics* (New York: McGraw-Hill)

Sprackling, M. (1991). *Thermal Physics* (Basingstoke and London: Macmillan)

Wilks, J. (1961). *The Third Law of Thermodynamics* (London: Oxford University Press)

Zemansky, M. W. (1957). *Heat and Thermodynamics*, fourth edition (New York: McGraw-Hill)

Texts on low-temperature physics

Mackinnon, L. (1966). *Experimental Physics at Low Temperatures* (Detroit: Wayne State University Press)

McClintock, M. (1964). *Cryogenics* (New York: Reinhold)

McClintock, P. V. E., Meredith, D. J. and Wigmore, J. K. (1984). *Matter at Low Temperatures* (London: Blackie)

2

Properties of solids at low temperatures

It may, at first, appear strange to choose to consider the properties of matter at low temperatures before explaining how the low temperatures are achieved, or how the experiments to determine those properties are performed. However, knowing the physical properties of the substances used in the attainment of low temperatures is central to the understanding of how the cooling processes work. It's a question of what comes first—the chicken or the egg?

2.1 General properties

Upon cooling to low temperatures, all gases eventually condense. Further cooling towards 0 K almost invariably leads to solidification, liquid helium being the only exception to this rule. These changes in state can be ascribed to the slowing down of atomic and molecular motions. As the thermal kinetic energy is reduced, the normally rather weak van der Waals intermolecular forces are eventually able to dominate. We observe an ordering or lowering in entropy of the material as it is cooled; see Section 1.3. The rather interesting permanent liquids, liquid ^{4}He and its lighter isotope ^{3}He, will be discussed in more detail in the next chapter. For now, we will consider the general and mechanical properties of common solids, paying particular attention to those used in cryogenic applications.

2.1.1 Mechanical properties

Solid materials that are flexible at room temperatures, like plastics and rubber, become hard and brittle as they are cooled. These polymer materials consist of long, chain-like molecules held together (crosslinked) by van der Waals forces. When stressed, the molecules uncoil and slip over one another, giving the materials their characteristic ductility. The ease with which they do this is related to the thermal motion of the molecules. At low temperatures, the attractive forces outbalance the thermal motions and the material strongly resists deformation, eventually becoming brittle. For this reason, the vast majority of plastics materials are useless at low temperatures. There is usually enough internal stress set up by thermal contraction during cooling to shatter the material, even in the absence of any external forces. One polymer which does retain some plasticity, even at liquid-helium temperatures, is polytetrafluoroethylene (PTFE), which consequently has a number of cryogenic uses, for instance where it is necessary to insulate electrical leads at low temperatures.

Metals become less ductile at low temperatures, though the amount of change depends on the crystal structure of the metal. Materials like aluminium, chromium–nickel stainless steel, copper and its common alloys, all with face-centred cubic structures, remain ductile at low temperatures. Iron and carbon steel, with body-centred cubic structures, become brittle upon cooling. The ductility in metals is due to the motion of crystallographic imperfections, known as dislocations, which allows the planes of atoms to slip over one another at much lower levels of applied stress than in perfect crystals. At high temperatures some atoms gain enough thermal energy to escape their lattice sites. The resulting vacancies and interstitials seem to assist the motion of dislocations, although the precise nature of the interactions is complicated. At low temperatures there are fewer thermally created point defects, and consequently the motion of dislocations is inhibited. The differences between metals of different crystal structure are related to the number of slip planes, that is planes of atoms that can slip relatively easily over one another, which in turn depends on the symmetry of the lattice.

The elasticity or 'springiness' of metals is generally affected only slightly by cooling. Changes in Young's modulus of the order 5–10 per cent are observed upon cooling from room temperature to liquid-helium temperatures. Stainless steel, for example, has a Young's modulus of 190 GN m^{-2} at room temperature and 200 GN m^{-2} at 4.2 K. The elastic properties are due to the spring-like nature of the forces holding the atoms together in the metal. In most metals the strength of these forces is not strongly temperature-dependent in the range of temperatures of interest to us. At much higher temperatures, violent thermal motion of the atoms can slightly weaken the forces, and so metals are found to strain more easily.

Elasticity in polymer materials is due to the uncoiling of the long chain molecules against the van der Waals forces which tend to restore the material to its unstrained state. The loss of elasticity in these materials at low temperatures, is related to the onset of brittle behaviour.

Most materials become stronger at low temperatures. The yield stress is a measure of how much stress the material can stand before deforming permanently, and upon cooling it is found to increase by up to two times in metals and of the order five to ten times in plastics. The ultimate stress or breaking point stress is more or less equal to the yield stress in brittle materials, but is much greater in ductile materials, which can deform to redistribute the stress. Generally, metals like stainless steel, aluminium and copper are all suitable for use at low temperatures. The scale of laboratory apparatus is such that the structural elements are rarely expected to be operated anywhere near their yield stress; the choice of metal has more to do with such factors as thermal and electrical conductivity than the mechanical properties, as we shall see later.

2.1.2 Thermal contraction

It is well known that materials change their dimensions as they cool. In nearly all cases, this change takes the form of thermal contraction, although a few exceptional materials actually expand, e.g. fused silica glass. In the case of metals, contraction can be up to one per cent in volume and nearly all takes place between room temperature and liquid-nitrogen temperature, 77 K. The atoms in metals are held by forces which can be likened to springs linking the atoms. A low-temperature theory due to Debye (Section 2.1.4) assumes that these forces are harmonic. At a finite temperature, the vibrating atoms behave as simple harmonic oscillators, vibrating with small amplitude symmetrically about their equilibrium positions. At high temperatures the vibrations become large, and so anharmonic effects in the restoring forces can become important, which gives rise to a slight increase in the mean separation of the vibrating atoms, resulting in the expansion of the material. In the case of plastics and rubber, the slowing of the thermal motion allows the molecules to be pulled closer together by the van der Waals forces. This causes a relatively large change in the dimension of the material (sometimes by as much as 30 per cent between room temperature and 4.2 K). Figure 2.1 shows the thermal contraction for a number of materials in common cryogenic use. The thermal contraction is highly non-linear. As stated earlier, very little contraction occurs at temperatures much below 77 K. This is consistent with thermodynamic arguments, which indicate that the coefficient of thermal contraction should approach zero as temperature approaches absolute zero.

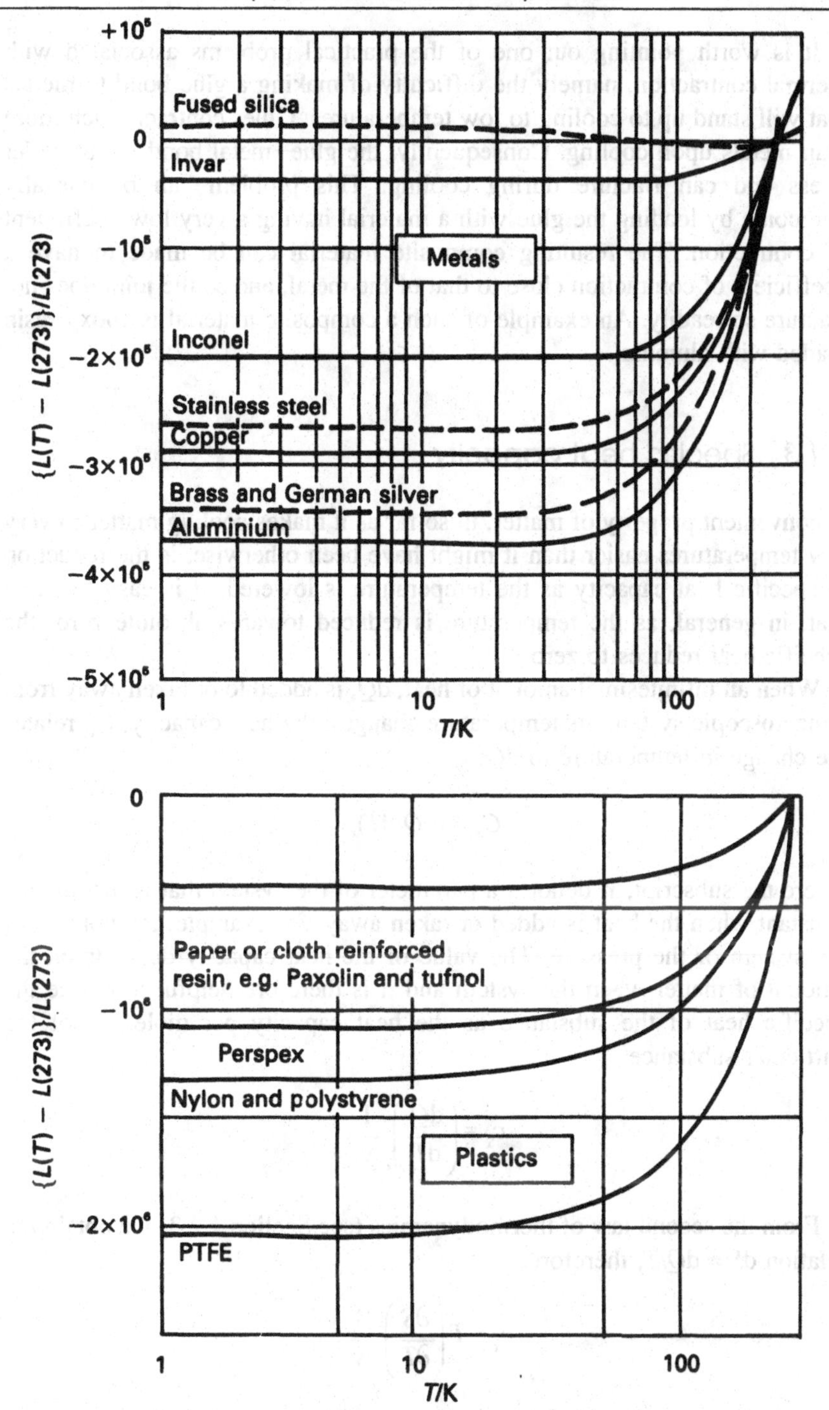

Figure 2.1 *Thermal expansion of materials in common cryogenic use [after Rose-Innes, A. C. (1973).* Low Temperature Laboratory Techniques *(London: The English Universities Press)]*

It is worth pointing out one of the practical problems associated with thermal contraction, namely the difficulty of making a glue bond to metals that will stand up to cooling to low temperatures. Glues contract much more than metals upon cooling. Consequently, the glue–metal bond is put under stress and can fracture during cooling. This problem can be partially overcome by loading the glue with a material having a very low coefficient of contraction. The resulting composite material can be made to have a coefficient of contraction close to that of the metal, and so the joint does not fracture so readily. An example of such a composite material is epoxy resin loaded with alumina.

2.1.3 Specific heat capacity

A convenient property of matter, in so far as it makes cooling matter to very low temperatures easier than it might have been otherwise, is the reduction in specific heat capacity as the temperature is lowered. It is easy to show that, in general, as the temperature is reduced towards absolute zero, the specific heat reduces to zero.

When an infinitesimal amount of heat, dQ, is added to or taken away from a macroscopic system, its temperature changes; the heat capacity, C_x, relates the change in temperature to dQ:

$$C_x = (\mathrm{d}Q/\mathrm{d}T)_x$$

where the subscript, x, denotes a parameter of the system that is maintained constant when the heat is added or taken away, for example, the volume of the system or the pressure. The value of the heat capacity depends on the amount of matter, ν, in the system and it is therefore helpful to define the specific heat of the substance as the heat capacity per mole, c_x, of the particular substance

$$c_x = \left(\frac{\mathrm{d}Q}{\mathrm{d}T}\right)_x \frac{1}{\nu}.$$

From the second law of thermodynamics (see Section 1.2.3) we obtain the relation dS = dQ/T, therefore

$$c_x = T\left(\frac{\partial S}{\partial T}\right)_x \frac{1}{\nu},$$

but the third law (Section 1.2.4) asserts that, as the temperature approaches absolute zero, the entropy tends to a constant value, S_0. Therefore, dS/dT = 0 and so $c_x = 0$ at absolute zero.

The value of the specific heat depends on the microscopic nature of the system under consideration and the nature of the thermal excitations within it. As an example, consider a monatomic ideal gas. It has, according to the equipartition theorem, an internal energy of $3RT/2$ joules per mole. The gas can do work, $p\,dV$, in expanding. Applying the first law of thermodynamics gives the heat absorbed in an infinitesimal process involving an ideal gas,

$$dQ = dE + p\,dV.$$

Therefore,

$$dQ = \frac{3R}{2}dT + p\,dV.$$

At constant volume $dV = 0$ and so

$$c_V = \left(\frac{dQ}{dT}\right)_V = \frac{3R}{2}\ \mathrm{J\,K^{-1}\,mol^{-1}}.$$

To find the specific heat at constant pressure, we use the equation of state of a mole of ideal gas, $pV = RT$, therefore,

$$dQ = \frac{3R}{2}dT + R\,dT,$$

and hence

$$c_p = \left(\frac{dQ}{dT}\right)_p = \frac{5R}{2}\ \mathrm{J\,K^{-1}\,mol^{-1}}.$$

In the case of diatomic gases, there are rotational and vibrational degrees of freedom associated with the molecules. To each of these there can be apportioned $RT/2$ Joules of internal energy, which increases the specific heat capacity by the appropriate amounts.

Although these results might seem to contradict the rule that the specific heat capacity tends to zero as the temperature approaches absolute zero, it must be remembered that, as the gas is cooled towards absolute zero, purely classical concepts, such as those used above, break down. Under these circumstances it is necessary to adopt a quantum-mechanical description of matter. The specific heat capacity of liquid helium, in its quantum-mechanical regime, will be considered in the next chapter.

The situation in solids is rather different; the internal energy exists as thermal vibrations of the lattice, and in the case of metals there is an additional contribution from the gas of free conduction electrons.

Considering now the contribution from lattice vibrations, each atom in a simple solid element is free to vibrate in three dimensions. For small displacements the restoring forces are harmonic, so a solid containing N atoms can be thought of as consisting of $3N$ independent one-dimensional harmonic oscillators, each of energy

$$E = \frac{p^2}{2m} + \frac{m^2 q^2}{2}.$$

The first term is the kinetic energy and the second the potential energy of the atomic oscillator. According to the equipartition theorem, an average thermal energy $kT/2$ can be ascribed to each of these. So, classically, the total vibrational energy is always equal to

$$E = 3NkT,$$

or, for one mole of the solid,

$$E = 3RT.$$

The specific heat capacity at constant volume (solids only change volume slightly due to expansion, so to within a few per cent $c_p = c_V$) is, at sufficiently high temperatures,

$$c_V = \left(\frac{\mathrm{d}E}{\mathrm{d}T}\right)_V = 3R,$$

or 25 J $\mathrm{K^{-1}mol^{-1}}$. This result is known as the law of Dulong and Petit and has been verified experimentally for a wide range of solids at high temperatures. This does not hold at lower temperatures; in fact, it is known that the specific heat must go to zero as $T \rightarrow 0$. At much lower vibrational energies, the harmonic oscillators must be considered as quantum-mechanical oscillators, not classical oscillators.

2.1.4 Phonons and the Debye theory of the specific heat

The first attempt at calculating the specific heat of a solid consisting of an assembly of quantum oscillators was made by Einstein. Central to his theory is the assumption that all the atoms vibrate independently at the same frequency. The theory was successful in that, at high temperatures, $c_V = 3R$, in agreement with the classical theory, and as T tends to zero, c_V also approaches zero, although rather more quickly than was observed in practice. Instead of treating the atomic oscillators as independent, Debye assumed that the solid could be treated as an elastic continuum. This assumption is reasonable if the wavelength of the normal modes of oscillation is greater than the interatomic spacing. This is indeed the case in most crystalline materials for vibrational frequencies up to about 10^{13} Hz.

A quantised harmonic oscillator with energy levels $E_n = (n + \frac{1}{2})\hbar\omega$ exchanges energy with its surroundings in units of $\hbar\omega$. These quanta of lattice vibration energy can be treated as quasi-particles called phonons. Like photons, they possess a momentum $p = \hbar k$ and travel with a velocity $v = \omega/k$ (where $k = 2\pi/\lambda$ is the wavenumber), but they are confined to exist only within the bounds of the crystal. The number of possible modes of oscillation of the lattice, having frequencies in the narrow interval between ω and $\omega + d\omega$, is given by $D(\omega)\,d\omega$, where $D(\omega)$ is called the phonon density of states. Theoretical analysis of the low-frequency vibrational modes of a three-dimensional lattice of volume V (see, for example, Kittel, 1976) gives

$$D(\omega) = \frac{3\omega^2 V}{2\pi^2 \bar{v}^3},$$

where the velocity, $\bar{v}$, represents an average over the longitudinal and two transverse modes of oscillation. This form for $D(\omega)$ is found to be a reasonable approximation to the actual density of states, at least at low frequencies, in a number of materials. For an example in copper, see Figure 2.2.

The crystal cannot, of course, support an infinite number of phonons; it is reasonable to assume that there are a maximum of $3N$ modes of oscillation available, as in the classical theory. When heat is added to a crystal, volume V, initially at absolute zero, the available phonon modes fill up until eventually all $3N$ are occupied. Because $D(\omega)$ is a monotonically increasing function of frequency, it must imply a cut-off frequency for the lattice, defined by

$$3N = \int_0^{\omega_D} D(\omega)\,d\omega.$$

Using the above form for $D(\omega)$ gives

$$\omega_D = \left(6\pi^2 \bar{v}^3 \frac{N}{V}\right)^{1/3}.$$

This is known as the Debye frequency of the lattice. It turns out to be approximately equal to the frequency at which the phonon wavelength is comparable to the interatomic spacing.

Phonons have zero-spin quantum number. They belong to a family of particles and excitations known as bosons and so, like ^{4}He atoms, they obey Bose–Einstein statistics (see Section 3.1). The probability of a phonon state of energy $E = \hbar\omega$ being occupied when the lattice is in thermal equilibrium at temperature T is given by

$$P(E) = \frac{1}{\exp(\hbar\omega/kT) - 1}.$$

It can be seen that, as T becomes large, this tends towards the classical Boltzmann factor, $\exp(-\hbar\omega/kT)$. The total energy, U, of the lattice at temperature T is the sum of the energies of all the modes present, plus, of course, the zero point energy, U_0,

$$U = U_0 + \int_0^{\omega_D} d\omega \frac{D(\omega)\hbar\omega}{\exp(\hbar\omega/kT) - 1}.$$

Evaluating the integral in the high-temperature limit, $kT \gg \hbar\omega_D$, gives

$$U = U_0 + 3NkT,$$

and so the heat capacity,

$$C_V = \left(\frac{dU}{dT}\right)_V = 3Nk.$$

For one mole of the substance, c_V is equal to $3R$, as given by the classical law of Dulong and Petit. Evaluating the integral in the low temperature limit, $kT \ll \hbar\omega_D$, gives

$$U = U_0 + \left(\frac{3N\hbar\omega_D}{5}\right)\left(\frac{\pi kT}{\hbar\omega_D}\right)^4;$$

therefore

$$c_V = \left(\frac{12\pi^4 R}{5}\right)\left(\frac{T}{\Theta_D}\right)^3 \ \text{JK}^{-1}\text{mol}^{-1},$$

where $\Theta_D = \hbar\omega_D/k$ is known as the Debye temperature. As T tends to zero, c_V is proportional to T^3, as observed in practice, with c_V falling to zero at absolute zero temperature. Despite the approximations inherent in the calculation of $D(\omega)$, the specific heat predicted by the Debye theory is in surprisingly good agreement with experimental measurements (Figure 2.3). Table 2.1 gives values of the Debye temperature for a variety of materials. In non-crystalline solids, the phonon density of states is much more difficult to calculate; however, the Debye theory still proves to be useful in a number of cases.

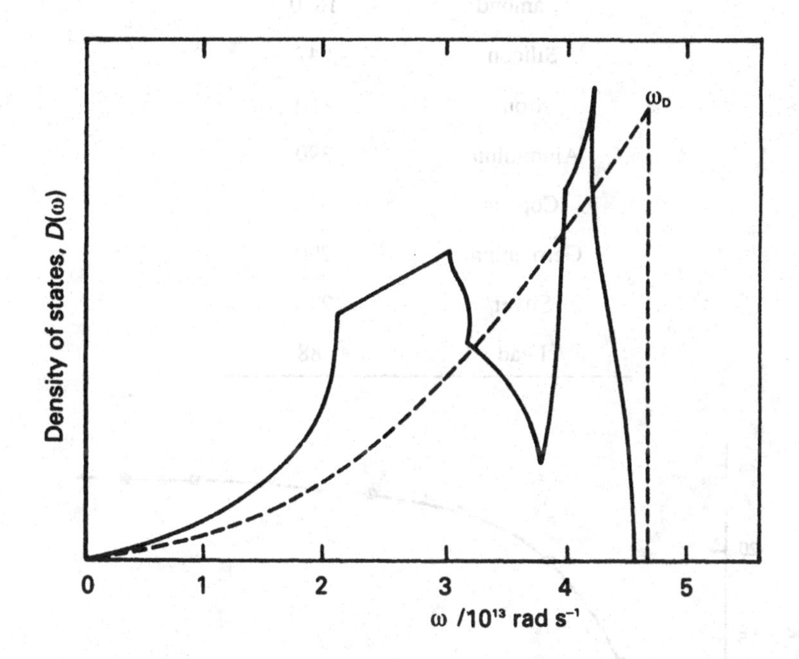

Figure 2.2 *Density of states for phonons in copper predicted by the Debye approximation. The broken line is the experimental density of states as deduced from neutron scattering experiments [after Svensson, E. C., Brockhouse, B. N. and Rowe, K. J. (1967).* Phys. Rev., *155, 619]. The area under the two curves is the same*

In metals, phonons make the most significant contribution to the specific heat, except at the very lowest temperatures, where the electronic contribution can become important (see Section 2.2).

2.1.5 Thermal conduction

Of great importance in low-temperature systems is the transport of heat by thermal conduction in the system. It is by conduction that parts of the system are cooled and brought into thermal equilibrium. Conduction is also partly responsible for unwanted heat leaks into low-temperature apparatus.

Table 2.1 Values of the Debye temperature for various crystalline solids [after Dekker (1970)].

Material	Θ_D /K
Diamond	1860
Silicon	647
Iron	420
Aluminium	390
Copper	315
Germanium	290
Silver	215
Lead	88

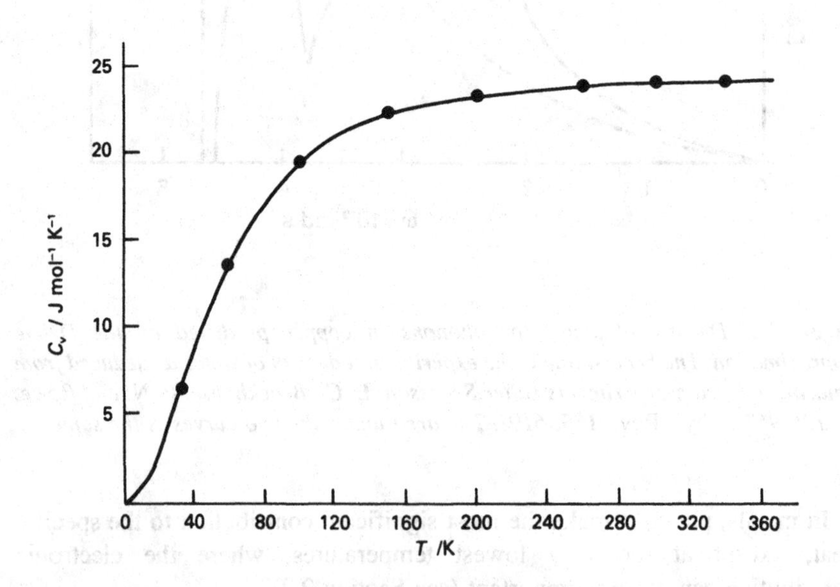

Figure 2.3 *Specific heat of silver, experimental points and the result of the Debye calculation (solid line) [after Dekker (1970)]*

Consider a long sample of material, of constant cross-sectional area A, with a temperature gradient dT/dx along it. Heat will flow along the sample from the higher to the lower temperature, bringing the sample towards thermal equilibrium. The heat flux $\dot{Q}$ along the sample is proportional to the temperature gradient, provided the latter is not too large:

$$\dot{Q} = \mathrm{K}A\,\frac{\mathrm{d}T}{\mathrm{d}x}.$$

The constant K is the coefficient of thermal conduction. In a gas, the heat is carried by the molecules themselves. The hotter, more energetic molecules intermingle with the cooler ones, bringing the whole system into equilibrium. In a solid the atoms cannot move far from their equilibrium positions and so the energy is carried by the gas of phonons. In metals the gas of free electrons is very effective in the transport of heat. For a 'gas' of n particles per unit volume, each having a heat capacity c, travelling with a mean velocity $\bar{v}$ and travelling a mean distance l before colliding with something, simple kinetic theory arguments give

$$\mathrm{K} = \frac{1}{3}n\bar{v}c\bar{l}.$$

This result is general; it applies whatever the heat carrier is. For a simple monatomic gas, collisions are between the gas molecules themselves; l is therefore inversely proportional to the number of molecules in the container. The mean velocity is proportional to $T^{1/2}$, and so therefore is K. A rigorous calculation, under the assumption that the gas atoms are hard spheres of mass m and diameter d gives

$$\mathrm{K} = \frac{75k}{64\sigma_0}\left(\frac{\pi kT}{m}\right)^{1/2},$$

where $\sigma_0 = \pi d^2$ is the collision cross section of the atoms.

In insulating solids, it is the gas of phonons that carries the heat. The number of phonons depends strongly on temperature; at low temperatures it is proportional to T^3 (this follows from the Debye theory; see Section 2.1.4). Phonons travel at the velocity of sound in the lattice, which is independent of temperature. The heat capacity of a phonon is of the order k and independent of temperature. The temperature dependence of the mean free path l is rather more complicated. At very low temperatures, phonons can travel long distances in crystals and are scattered by the boundaries or fixed imperfections. These scattering processes do not depend on temperature, and so K is proportional to T^3 at low temperatures, as observed in practice. At higher temperatures it is generally found that K is proportional to T^n where $n < 3$, as temperature-dependent scattering processes, like phonon-phonon scattering, become important in determining l.

In metals there is an additional contribution from the free electrons. At low temperatures this contribution is dominant and will be considered in the next section.

2.2 Electronic properties of materials at low temperatures

Materials that are electrically insulating at room temperature remain so upon cooling. On the other hand, metals and semiconductors exhibit some interesting phenomena at low temperatures, which are related to the behaviour of the gas of free electrons found in these materials. Furthermore, the electronic properties of these materials are exploited in low-temperature experiments, for example, in the measurement of temperature. These properties have been and are still the subject of considerable research. Only a brief outline of the most basic properties is given here.

2.2.1 Electrical conductivity of metals

In a metal, at least one electron from every atom is able to move throughout the crystal, which could be visualised as a periodic array of positive ions, embedded in a sea of free electrons. The periodic modulation of the potential felt by an electron, as it moves in the lattice of positive ions, acts to quantise the electron energies. Electrons have a spin quantum number of ½, they are known as fermions and obey the Pauli exclusion principle, which is that only one electron can occupy a particular quantum state. The electronic quantum states are specified both by the energy level and the electron spin. This means that only two electrons, one with spin state +½ and the other with −½, can occupy each energy level. Now, imagine a metal stripped of all its electrons, at absolute zero temperature; as electrons are added, they go in pairs into the lowest available energy states, until all the electrons have been accommodated. The highest energy level occupied is called the Fermi level, E_F (see Figure 2.4). For a metal with n electrons per unit volume, $E_F = (\hbar^2/2m)\{3\pi^2 n\}^{2/3}$, where m is the electron mass. At a finite temperature, the electrons possess some thermal energy. This leads to excitation of electrons below and within kT of the Fermi energy, to energy levels just above the Fermi level. It is these electrons, within kT of the Fermi energy, that are responsible for the thermal and electronic properties of the metal. Mathematically, the probability of an electronic energy level, of energy E, being occupied can be written

$$P(E) = \left\{ 1 + \exp\left(\frac{E - E_F}{kT} \right) \right\}^{-1} .$$

For $kT \ll E_F$, this expression approximates closely to the zero temperature situation and the electrons form what is known as a degenerate Fermi gas.

This is the case for most metals, even at room temperature, since the temperature at which the electronic thermal energy is comparable to the Fermi energy is E_F/k and of the order 10^5 K. The temperature E_F/k is called the Fermi temperature of the system.

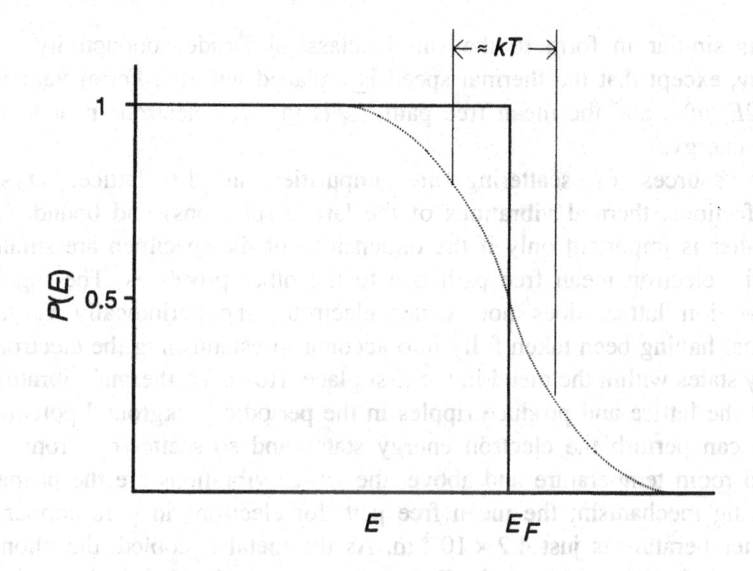

Figure 2.4 *The Fermi Function, at absolute zero (solid line) and at a finite temperature,* T *(broken line)*

When a potential difference is applied across a metal, the electrons are accelerated by the electric field within the metal and an electric current flows. The theory of conduction by a degenerate electron gas, worked out by Sommerfeld, treats the effect of the applied field and the scattering of electrons as perturbations of the electronic energy distribution. The scattering impedes the electron motion and gives rise to the resistance of the metal. If the perturbations are small, then only the electrons within kT of the Fermi energy need to be considered. In order to comply with the Pauli exclusion principle, electrons in states well below the Fermi level cannot undergo any scattering, because all the states close to their initial energy are fully occupied, i.e. there are no available states for the electrons to scatter into. Sommerfeld's model leads to the following expression for the electrical conductivity:

$$\sigma = \frac{ne^2 \bar{l}_F}{mv_f}.$$

This is similar in form to the simple classical Drude conductivity, $\sigma = ne^2\bar{l}/m\bar{v}$, except that the thermal speed is replaced with the Fermi velocity, $v_F = (2E_F/m)^{1/2}$, and the mean free path, l_F, is that for electrons near to the Fermi energy.

The sources of scattering are impurities in the lattice, crystal imperfections, thermal vibrations of the lattice (phonons) and boundaries. The latter is important only if the dimensions of the specimen are smaller than the electron mean free path due to the other processes. The regular positive ion lattice does not scatter electrons, the periodically varying potential having been taken fully into account in establishing the electronic energy states within the metal in the first place. However, thermal vibrations distort the lattice and produce ripples in the periodic background potential, which can perturb the electron energy states and so scatter electrons. At around room temperature and above, the lattice vibrations are the primary scattering mechanism; the mean free path for electrons in pure copper at room temperature is just 4.2×10^{-8} m. As the metal is cooled, the phonon population decreases. Above the Debye temperature, the electrical resistivity due to phonon scattering, ρ_{pho}, is approximately proportional to the temperature, T. Below the Debye temperature the phonon population decreases as T^3, but the effectiveness of the lower energy phonons in scattering electrons is rather limited, so at low temperatures ρ_{pho} is found to be approximately proportional to T^5. In most metals the resistivity does not go on decreasing to zero as the temperature approaches absolute zero, but reaches some limiting value, ρ_{imp}, which is due to temperature-independent scattering by fixed defects and impurities. The total electrical resistivity can be expressed in terms of the two contributions ρ_{pho} and ρ_{imp}, according to Mathiessen's rule:

$$\rho = \rho_{imp} + \rho_{pho}.$$

Figure 2.5 shows the electrical resistivity of copper samples of varying purity. The transition from phonon-dominated scattering to impurity scattering, at temperatures around 10 K, is clearly seen, the change-over temperature being lower in the pure samples.

There are a number of metals, known as superconductors, that make a transition to a zero resistance state as the temperature is reduced below some critical value. This is due not to their being highly pure, but to other effects which will be considered later, in Section 2.3.

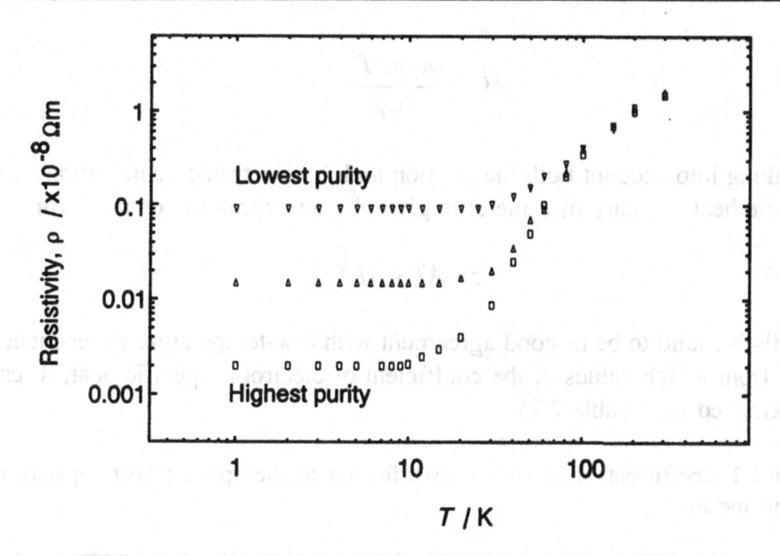

Figure 2.5 *Temperature dependence of the electrical resistivity of samples of copper with differing purity*

2.2.2 Electronic specific heat and thermal conduction

The electron gas makes a very significant contribution to the thermal conductivity of metals; it is far more important than the phonon gas at all the temperatures of interest. Metals are characterised by their high thermal conductivities. On the other hand, the electronic contribution to the specific heat is important only at the very lowest temperatures. In both cases, it is only the electrons within about kT of the Fermi energy that need to be considered. To a first approximation and at sufficiently low temperatures ($kT \ll E_F$), the number of electrons affected is proportional to kT/E_F, that is, the ratio of the area under that part of the Fermi function within kT of the Fermi energy to the area under the entire curve, as shown in Figure 2.4. To each individual electron can be attributed an average heat capacity of $3k/2$, and so it is found that electronic specific heat is given by an expression of the form $c_e = 3nk^2T/2E_F$, where n is the number of free electrons per mole. It has already been shown that the phonon specific heat is proportional to T^3, and tends to zero much faster than the electronic specific heat as the temperature is reduced to 0 K. Therefore, below a certain (usually very low) temperature, the electronic term will dominate. A rigorous calculation of the electronic specific heat capacity (see for instance Blakemore, 1985) gives

$$c_e = \frac{\pi^2 nk^2 T}{2E_F}.$$

Taking into account both the phonon and the electronic contributions, the specific heat capacity of a metal is given by an expression of the form

$$c = AT + BT^3.$$

This is found to be in good agreement with low-temperature experimental data, from which values of the coefficient of electronic specific heat, A, can be extracted (see Table 2.2).

Table 2.2 Coefficient of electronic contribution to the specific heat capacity of various metals

Metal	A J mol^{-1} K^{-2}
Iron	0.50×10^{-3}
Sodium	1.55×10^{-3}
Aluminium	0.91×10^{-3}
Copper	0.71×10^{-3}
Silver	0.64×10^{-3}

The thermal conductivity of the electron gas in a metal can be expressed in terms of c_e (see Section 2.1.5):

$$K = \tfrac{1}{3} n c_e v_F \bar{l}.$$

The Fermi velocity, v_F, is independent of temperature. The mean free path, $\bar{l}$, depends on temperature and three regions of different behaviour can be identified:

1 At very low temperatures, there are few phonons and the scattering is due to impurities and crystallographic imperfections. Such scattering is independent of temperature and so $K \propto T$.

2 At higher temperatures, but still below the Debye temperature, Θ_D, the electrons are scattered by phonons; $\bar{l}$ is then inversely proportional to the number of phonons, which for $T < \Theta_D$ increases as T^3, hence $K \propto T^{-2}$.

3 Above the Debye temperature, the mean free path of electrons due to phonon scattering is inversely proportional to T and so K is independent of temperature in this range.

If K_i and K_p are the electronic thermal conductivities corresponding to impurity scattering and phonon scattering respectively, then, in the normal situation when both scattering processes are working together, the resultant electronic thermal conductivity is given by

$$\frac{1}{K} = \frac{1}{K_i} + \frac{1}{K_p} .$$

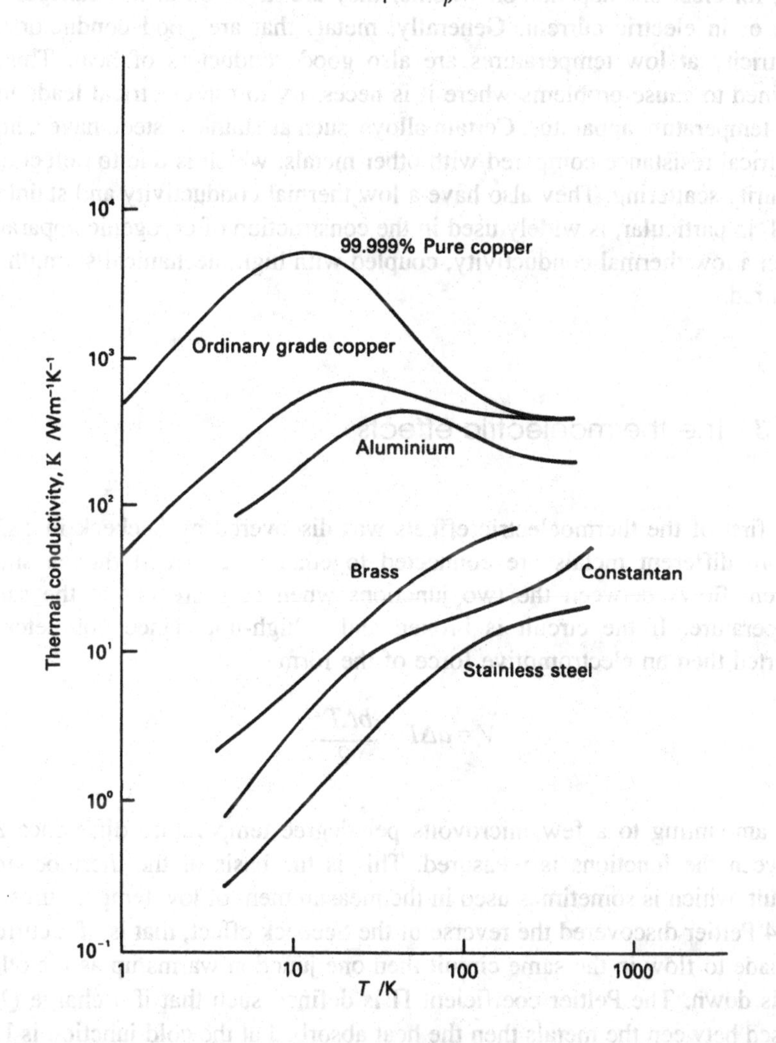

Figure 2.6 *Thermal conductivities of various metals versus temperature [after Rose-Innes, A. C. (1973).* Low Temperature Laboratory Techniques *(London: The English Universities Press)]*

Figure 2.6 shows the measured thermal conductivities of various metals commonly used for cryogenic purposes as a function of temperature.

Electrons are responsible for the transport of both heat and electric current in a metal. It is, therefore, not surprising that the electrical and thermal conductivities are related in some simple manner. At fairly high temperatures and certainly above the Debye temperature, they are related by the Wiedemann–Franz law, that is, $K/\sigma T = constant$. This is also true at very low temperatures where impurity scattering is dominant. In the middle range ($T \approx \Theta_D$) there is a departure from this law, which implies that the mean free path for electrons depends on whether they are involved in the transport of heat or in electric current. Generally, metals that are good conductors of electricity at low temperatures are also good conductors of heat. This is inclined to cause problems where it is necessary to run electrical leads into low-temperature apparatus. Certain alloys, such as stainless steel, have a high electrical resistance compared with other metals, which is due to defect and impurity scattering. They also have a low thermal conductivity and stainless steel, in particular, is widely used in the construction of cryogenic apparatus when a low thermal conductivity, coupled with high mechanical strength, is required.

2.2.3 The thermoelectric effects

The first of the thermoelectric effects was discovered by Seebeck in 1821. If two different metals are connected together in a circuit then a small current flows between the two junctions when they are not at the same temperature. If the circuit is broken and a high-impedance voltmeter is inserted then an electromotive force of the form

$$V = a\Delta T + \frac{b\Delta T^2}{2}$$

and amounting to a few microvolts per degree temperature difference ΔT between the junctions is measured. This is the basis of the thermocouple circuit, which is sometimes used in the measurement of low temperatures. In 1834 Peltier discovered the reverse of the Seebeck effect, that is, if a current is made to flow in the same circuit then one junction warms up as the other cools down. The Peltier coefficient Π is defined such that if a charge Q is passed between the metals then the heat absorbed at the cold junction is ΠQ Joules. The third and weakest of the thermoelectric effects is the Thomson heat: if a current flows down a wire along which a temperature gradient is maintained then there is a heating of the wire. The heating at any point along the length is due to the sum of three components: Joule or I^2R heating, conduction of heat from the hotter end, and the Thomson heat. The latter can lead to heating or cooling of the point, depending on the relative directions of the current flow and the thermal gradient. This is seen to arise as follows: The energy of the electrons is dependent on their temperature, with the electrons at the hot end having a higher energy. As they diffuse down the metal under the influence of the electric field, the electrons collide and transfer some of their heat to the lattice. Another contribution to the Seebeck and Peltier effects is the contact potential. When two different metals are connected together, a small potential, equal to the difference between the work functions of the two metals, appears across the junction. The work function, which is the minimum energy that has to be supplied for an electron to escape the metal, depends weakly on the temperature and so the contact potential is temperature-dependent. All the thermoelectric effects are interrelated thermodynamically and are conveniently expressed in terms of the thermopower S of the material

$$\frac{dS}{dT} = \frac{h_T}{T},$$

where h_T is the Thomson heat. The Seebeck e.m.f. is given in terms of the thermopower of the two metals S_1 and S_2 and the junction temperatures T_1 and T_2 by

$$V = \int_{T_1}^{T_2} (S_2 - S_1)\, dT,$$

and the Peltier coefficient by

$$\Pi = (S_2 - S_1)\, T.$$

So, in order to predict the temperature dependence of the thermoelectric effects, it is only necessary to know how the thermopower depends on temperature. In practice it is found to vary widely between similar metals. For example, in the case of Lithium at temperatures above 100 K, $S \approx +37$ nV K^{-1}, while for Sodium $S \approx -47$ nV K^{-1}. The thermopower is also very strongly dependent on the purity of the metal. For all materials the thermopower tends to zero at 0 K, although it does sometimes show a peak at temperatures below about 50 K owing to phonon drag, that is, the 'dragging along' of phonons by the electrons which transfer some of their momentum to the phonons in electron–phonon collisions. The phonons are unable to pass this momentum on to the crystal lattice as a whole, on the same timescale as the electron–phonon collisions.

2.2.4 Electrical conduction in semiconductors

Semiconductors are a technologically important class of materials. It is possible to modify the electronic properties of pure semiconductors by doping them with specific impurities and so make electronic devices, such as diodes and transistors. Unlike in a metal, at zero temperature, all the electrons in a semiconductor are bound to their parent atoms. However, the binding is much weaker than in an insulating material, and upon warming some electrons gain enough thermal energy to escape. It is these electrons and the 'holes' they leave behind that contribute to the conduction, or intrinsic conduction, as it is known. In pure silicon, the energy needed to release an electron is 1.1 eV which corresponds to a temperature of 12 000 K. So, pure silicon is an insulator at the temperatures we are considering. The effect of doping is to produce centres to which the electrons are more weakly bound. If we substitutionally add atoms of a donor that is a pentavalent element, e.g. phosphorus, to a concentration of, say, one phosphorus atom to every 10^8 silicon atoms then the electron not involved with bonding remains loosely attached to the donor centre. The resulting state is analogous to the hydrogen atom, except that the Bohr radius of the orbit is many atomic spacings. The ionisation energy of the donor state is of the order of a few millielectronvolts, so that a temperature of about 100 K is required to release the electron for conduction. The variation of free-electron concentration with temperature is as shown in Figure 2.7 (Smith, 1979). The free electrons obey Fermi statistics like those in a metal.

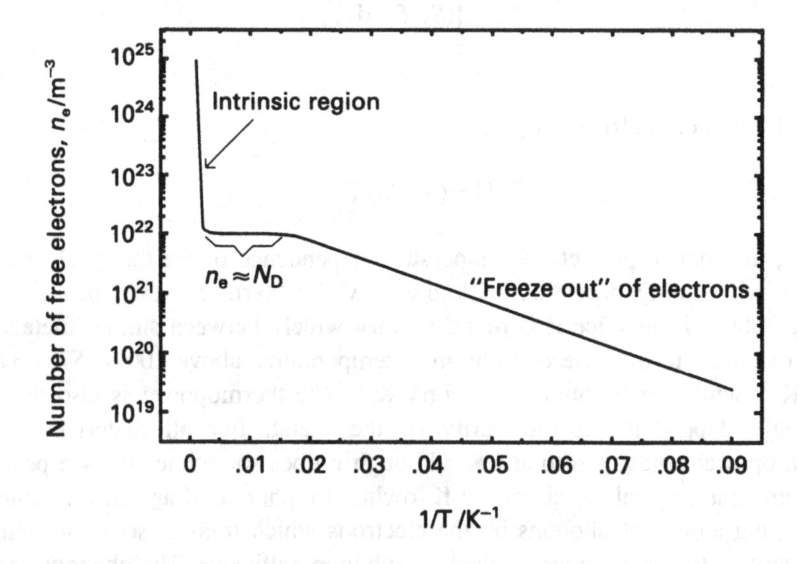

Figure 2.7 *Variation of conduction electron density with temperature for Germanium doped with Phosphorus donors to a concentration of 10^{22} m^{-3}*

However, because the electron concentration is smaller in semiconductors, the value of E_F is drastically reduced, and typically kT is comparable to E_F at temperatures of only 20 K or so. The electron gas is referred to as non-degenerate and many of the theoretical simplifications applicable to electrons in metals are not relevant to semiconductors, except at the very lowest temperatures. We will now consider the variation of electrical properties with temperature at temperatures below 300 K. It is customary to define a quantity called the mobility, μ, such that the conductivity of a semiconductor is given by

$$\sigma = ne\mu,$$

where n is the number of free electrons per cubic metre and e the electronic charge. At the higher temperatures, the electron concentration n is fairly constant, the mobility is controlled by the scattering of electrons by thermal phonons and is proportional to $T^{-3/2}$. As the temperature is reduced, scattering by impurities and defects becomes dominant. The scattering by neutral defects is independent of temperature, but the mobility due to scattering of electrons by charged defects, such as ionised donor atoms, varies approximately as $T^{3/2}$. This temperature dependence arises because the

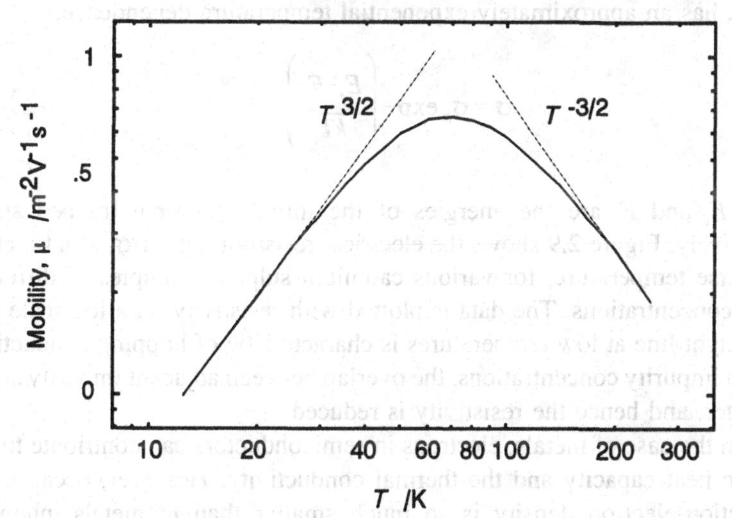

Figure 2.8 *Theoretical dependence on temperature of the drift mobility of electrons in a doped silicon sample*

'Rutherford-type' scattering depends on the electron energy (Blakemore, 1985). The overall mobility may be obtained by combining these two contributions in accordance with Mathiessen's rule. Figure 2.8 shows the expected temperature dependence of the mobility of a silicon sample with an ionised donor concentration of about 10^{21} m^{-3}. The mobility of a small number of samples has been observed to behave in this way, at least over a restricted range of temperature. However, it is more normal to find that, for a number of reasons, the electron or hole mobility depends on temperature in a rather different way to that predicted by such simple arguments. As the temperature is lowered still further, the electrons become trapped or 'localised' by the potential associated with impurities and the donor centres, so the number of free electrons *n* decreases and hence the conductivity falls. However, another electron transport mechanism becomes apparent in this regime, that of hopping conduction (Mott and Davis, 1979). Consider an electron that is trapped in the potential well of an impurity; this situation is analogous to the wave-mechanical problem of a particle in a finite-potential well. The electron's wavefunction extends outside the well and may overlap with the potential well of an adjacent, unoccupied impurity. There is a finite probability that the electron is able to 'tunnel' out of its potential well and hop to the adjacent impurity state which captures it. If there is a difference in the energies of the impurity states, then energy may have to be supplied to make the electron hop. The necessary energy is supplied by thermal phonons, giving rise to thermally activated hopping conduction. Likewise, if the final state is at a lower energy, then energy is released in the form of thermal phonons. Hopping or nearest-neighbour hopping conduction, as it is known, has an approximately exponential temperature dependence,

$$\sigma = \sigma_0 \exp -\left(\frac{E_f - E_i}{kT}\right),$$

where E_i and E_f are the energies of the initial and final trapped states respectively. Figure 2.9 shows the electrical resistivity, $\rho = 1/\sigma$, as a function of inverse temperature, for various cadmium sulphide samples of differing defect concentrations. The data is plotted with resistivity on a log scale and the straight line at low temperatures is characteristic of hopping conduction. At high impurity concentrations, the overlap between adjacent impurity states is greater, and hence the resistivity is reduced.

As in the case of metals, electrons in semiconductors can contribute to the specific heat capacity and the thermal conductivity. However, because the conduction-electron density is so much smaller than in metals, phonons dominate these quantities in semiconductors. On the other hand, some of the thermoelectric effects can be very large, for example, the Peltier effect in some semiconductor junctions can be used as the basis of small refrigeration units.

There is still a considerable research effort into the low-temperature properties of semiconductors. Particularly exciting is the class of device in which size effects, in one or more dimensions, are important. One such device is the silicon metal-oxide semiconductor field-effect transistor

(MOSFET), as shown in Figure 2.10, in which the electrons are confined to a very narrow sheet within 2–3 nm of the Si–SiO_2 interface by means of an electric potential applied perpendicular to the interface. The potential is produced by positively biasing a metal gate on top of the SiO_2 insulating layer. Like particles confined in a narrow well, the electron energy levels corresponding to motion perpendicular to the interface are quantized. If the temperature is sufficiently low ($T < 100$ K), and provided that there are not too many electrons, they are all confined in the lowest quantum level. They are, however, still able to move about freely in the plane parallel to the interface and form what is known as a two-dimensional electron gas.

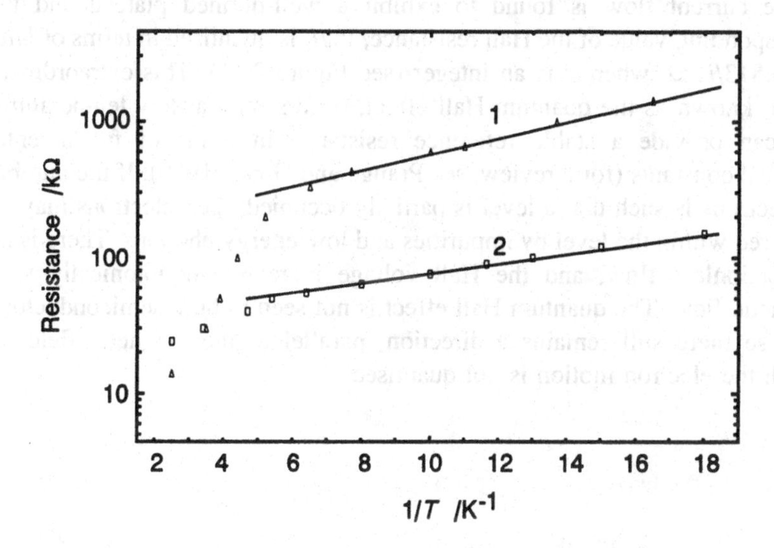

Figure 2.9 *Low temperature resistivity of samples of cadmium sulphide with differing types of defect. The linear dependence on inverse temperature at low temperatures is characteristic of hopping conduction, the activation energy being proportional to the slope of the line.*

When a magnetic field, B, is applied in a direction normal to the plane of the two-dimensional gas, the electrons make cyclotron orbits in the plane. The cyclotron orbits are quantised and the corresponding energy bands are called Landau levels; they are centred on energies $E_n = (n + \frac{1}{2})\hbar\omega_c$, where n is an integer, and the cyclotron frequency, $\omega_c = eB/m^*$, where m^* is the effective electron mass. The degeneracy of each level, that is the number of electrons it is able to accommodate, is given by $eB/\hbar$. Under conditions of sufficiently high magnetic field ($\mu B \gg 1$), the time taken by an electron to complete a cyclotron orbit is less than the mean scattering time for electrons and the Landau levels are narrow and clearly separated. Now suppose the total number of electrons present in the 2-D layer, which can be varied by changing the gate bias voltage, is such that all of the Landau levels, up to and including any particular one, are completely full of electrons and the next highest is empty. The absence of any empty states in the full Landau levels means that scattering of electrons is forbidden by the Pauli exclusion principle (if the electron is to be scattered then there must be an unoccupied quantum state for it to scatter into). If the temperature is low then scattering of electrons from within the occupied Landau levels to the empty ones at higher energy is not possible, either (the separation of adjacent Landau levels in a silicon MOSFET at a field of 5T is 3.5 meV, in temperature terms about 40 K). The current then flows without dissipation and the voltage parallel to the current flow direction drops to zero. The Hall voltage, V_H, perpendicular to the current flow is found to exhibit a well-defined plateau and the corresponding value of the Hall resistance, V_H/I, is quantised in terms of h/ie^2 or $25.813/i$ kΩ, where i is an integer (see Figure 2.11). This extraordinary effect, known as the quantum Hall effect, occurs only at low temperatures and can provide a stable reference resistance in terms of fundamental physical constants (for a review, see Prange and Girvin 1990). If the number of electrons is such that a level is partially occupied, then electrons may be scattered within the level by impurities and low-energy phonons. There is no dissipationless flow, and the Hall voltage increases monotonically with magnetic field. The quantum Hall effect is not seen in bulk semiconductors, because there still remains a direction, parallel to the magnetic field, in which the electron motion is not quantised.

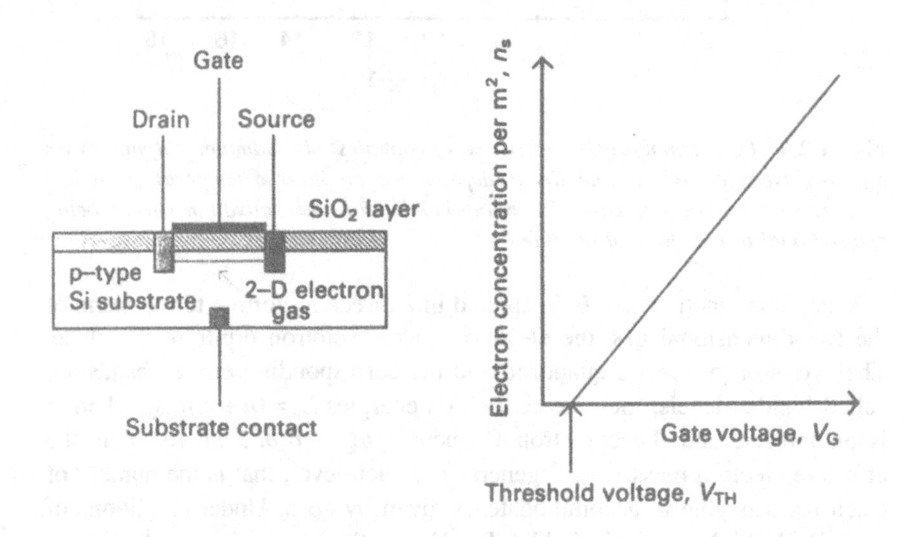

Figure 2.10 *A metal-oxide-semiconductor-field-effect-transistor structure and the variation of areal density of the 2-D electron gas with gate voltage*

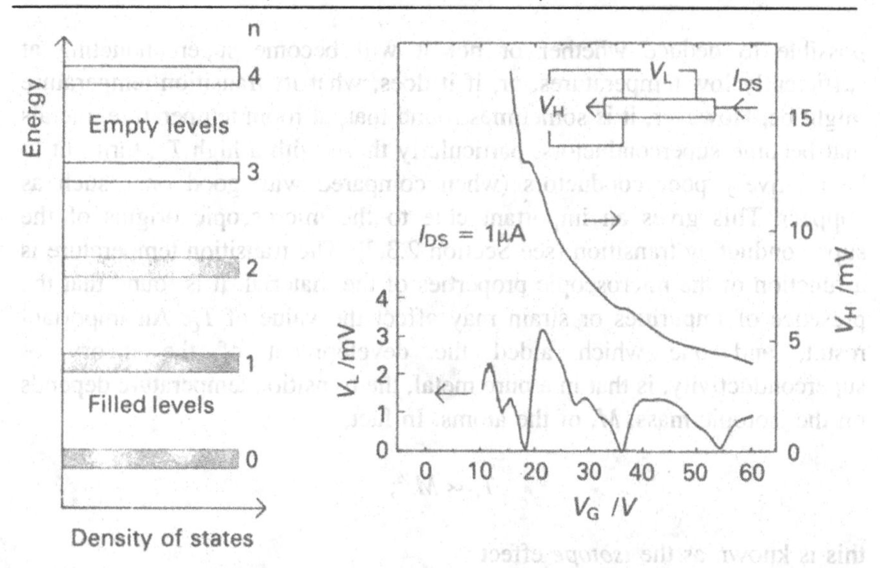

Figure 2.11 *(Left) Idealised density of states curve, showing the number of allowed electron quantum states as a function of energy, for 2-D electrons in a magnetic field. The states are broken up into bands called Landau levels. In this case, all of the available states in levels with n up to and including 2 are occupied with electrons and the higher levels are empty. (Right) Variation of longitudinal and Hall voltage with gate voltage,* V_G*, at constant drain—source current,* I_{DS}*. The plateaus in the Hall voltage correspond to quantised values of the Hall resistance,* V_H/I_{DS}*, given by* h/ie^2*. At these points, the longitudinal resistance drops to zero, i.e. the electrons drift down the length of the sample without scattering*

2.3 Superconductivity

It was mentioned briefly in Section 2.2.1 that below a certain critical temperature, some metals exhibit zero electrical resistivity to direct current (D.C.). This remarkable phenomenon, first discovered in mercury by Onnes in 1911, is called superconductivity and has been found to occur in more than 40 metals and alloys.

Figure 2.12 shows the dependence of resistance on temperature for a typical metallic superconductor. Above the transition temperature, T_C, the metal behaves in a perfectly normal manner, but at T_C the resistance falls rapidly to zero. The transition temperatures of various materials are listed in Table 2.3.

2.3.1 Physical properties of superconductors

On the basis of the properties of a material at high temperatures, it is not possible to deduce whether or not it will become superconducting at sufficiently low temperatures, or, if it does, what its transition temperature might be. However, it is sometimes found that, at room temperature, metals that become superconductors, particularly those with a high T_C, turn out to be relatively poor conductors (when compared with good ones such as copper). This gives an important clue to the microscopic origins of the superconducting transition (see Section 2.3.2). The transition temperature is a function of the microscopic properties of the material. It is found that the presence of impurities or strain may affect the value of T_C. An important result, and one which aided the development of the theory of superconductivity, is that in a pure metal, the transition temperature depends on the isotopic mass, M, of the atoms. In fact,

$$T_C \propto M^{½};$$

this is known as the *isotope* effect.

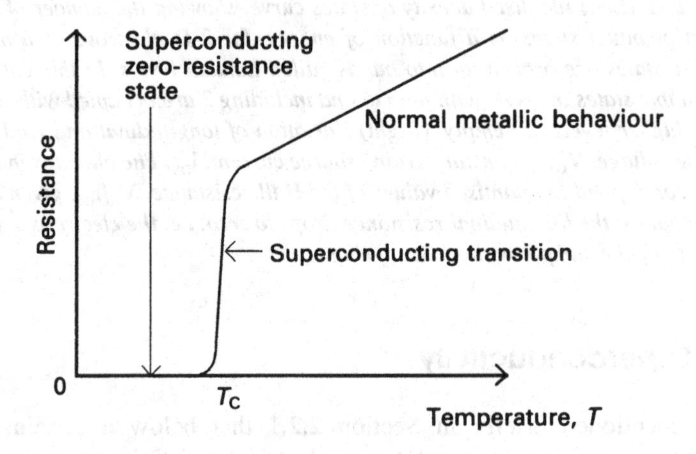

Figure 2.12 *Resistance–temperature characteristics of a typical metallic superconductor. In pure samples, the transition between normal and superconducting states can take place over less than 1 mK*

Measuring the decay of a persistent current in a closed ring of superconductor (Quinn and Ittner, 1962), enables an upper limit of 10^{-25} Ω m to be put on the value of the electrical resistivity, which is many orders of magnitude lower than the 4.2 K resistivity of the purest copper. At finite temperatures the resistivity presented to A.C. at all frequencies is non-zero, albeit very small. At high frequencies (> 10^9 Hz), superconductivity disappears altogether. By studying the infrared absorption as a function of frequency in a number of superconductors, it has been found that the zero resistance state breaks down at a frequency of about $f_C = 4kT_C/h$.

Table 2.3 Transition temperatures for a selection of superconductors.

Material	T_C / K
Niobium—Tin	18.5
Niobium—Titanium	16.0
Niobium—Gallium	14.5
Niobium	9.2
Lead	7.2
Mercury	4.12
Tin	3.71
Indium	3.37
Aluminium	1.18
Gallium	1.1
Zinc	0.85
Cadmium	0.54

The superconductive state is strongly influenced by magnetic fields. In fact, it is the particular response of a material to a magnetic field that classes it as a superconductor, as opposed to a normal metal with a vanishingly small electrical resistivity. Superconductors are perfect diamagnets; that is, if a superconductor is placed in a weak magnetic field and is cooled through its transition temperature then the magnetic flux is expelled from its interior. This is known as the Meissner effect. In the case of a superconducting hollow cylinder, cooling to below the transition temperature has the effect of trapping within the core of the cylinder any magnetic flux that was threading it prior to the transition.

Stronger magnetic fields are able to penetrate the bulk of the superconductor and quench the superconductivity. The flux density, B_C, at which this happens depends on temperature according to the approximate relation

$$B_C = B_0 \left\{ 1 - \left(\frac{T}{T_C} \right)^2 \right\},$$

where B_0 is the critical flux density at $T = 0$ (see Figure 2.13). Table 2.4 lists

B_0 for a few selected superconductors. One possible explanation for the Meissner effect is in terms of persistent supercurrents flowing in the surface of the superconductor (inner and outer surfaces in the case of the cylindrical sample).

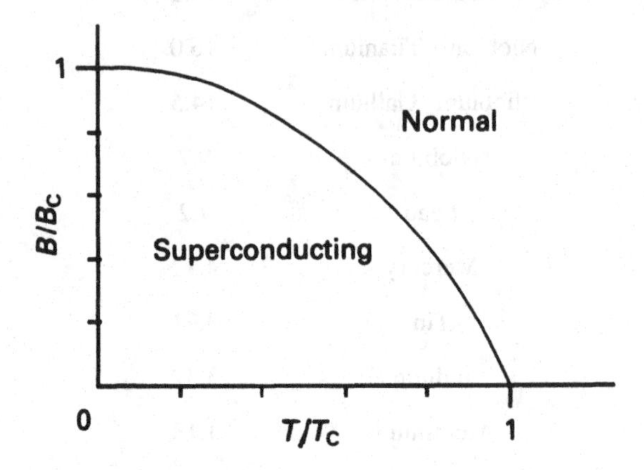

Figure 2.13 *Dependence on temperature of the magnetic flux density required to quench superconductivity.* B_0 *is the critical flux density at* $T = 0$

These give rise to a magnetic field, which exactly cancels the effect of the applied field. This explanation would imply that there is an infinite surface current density, unless the magnetic field is able to penetrate into the superconductor by a small amount. In fact, the field decays exponentially with depth, x, according to the equation

$$B(x) = B_a \exp -\left(\frac{x}{\lambda_L}\right).$$

Where B_a is the applied magnetic flux density and λ_L is called the London penetration depth, after the theoretical physicist F. London, who first studied this phenomenon. The penetration depth is typically of the order of tens of nanometres at $T = 0$ K. London also argued that, because superconductivity was a quantum-mechanical phenomenon, the persistent currents would be quantised. Therefore, the flux trapped in a superconducting cylinder would also be quantised. Flux quantisation has been observed experimentally in tiny superconducting rings (Deaver and Fairbank, 1961). The value of the flux quantum $\phi = h/2e$ implies that the current carriers possess a charge which is twice that on an electron $(-e)$. It should be noted that in the hypothetical case of a normal metal with infinite conductivity, it is not predicted that the field is excluded from the bulk. Upon becoming infinitely conducting, the normal metal traps within its bulk any magnetic field that was penetrating it prior to the transition and this cannot, subsequently, be changed by changing the external field. In other words, it becomes a perfect magnet.

A superconductor may be classified as one of two types, depending on the details of its response to an applied magnetic field. In the case of the so-called type-1 superconductors, provided the geometry is such that the field distribution around the sample is homogeneous, flux penetration into the bulk and quenching of the zero resistance state occurs discontinuously (Figure 2.14). Type-2 superconductors, on the other hand, allow some flux penetration above a lower critical flux density, B_{C1}, while still maintaining the zero-resistance state. Total flux penetration and quenching of the superconductivity often occurs at a much higher critical flux density, B_{C2}. In the region between the two critical flux densities the superconductor is in the so-called mixed state, and the flux penetrates the sample as a regular array of tubes or 'vortices'. The vortices have a radius equal to the penetration depth and the total flux through each is equal to one flux quantum. Type-2 superconducting materials are useful for making electromagnets which are capable of generating high magnetic fields with low power dissipation. An example of such a material is Niobium–Tin (Nb_3Sn), which has an upper critical field of 22 T at 4 K, some three orders of magnitude greater than the critical fields of the type-1 superconducting elements listed in Table 2.4.

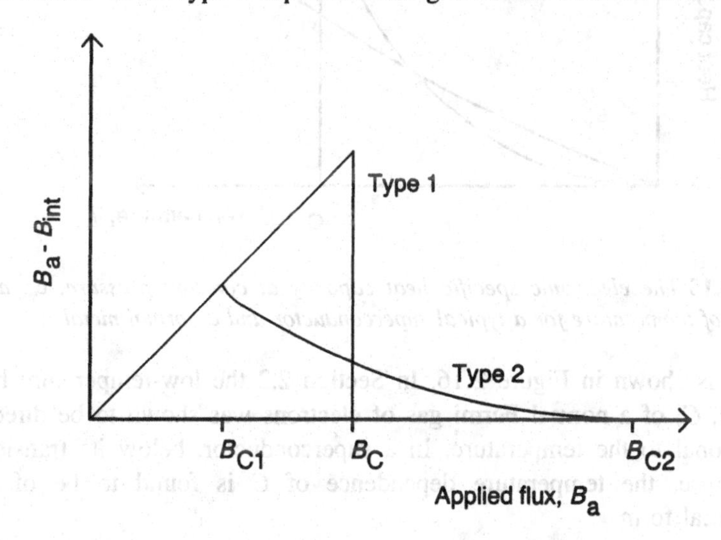

Figure 2.14 *The dependence of internal magnetic flux density,* B_{int}*, on applied flux density,* B_a*, in type-1 and type-2 superconductors*

The electronic specific heat capacity of a typical superconductor is shown in Figure 2.15. The anomaly at T_C is indicative of a second-order phase transition, that is, one not involving a heat of transformation of the electron system to a more ordered state. The corresponding decrease in the electronic

Table 2.4 Zero-temperature critical fields for a few superconducting elements

Element	B_0/T
Lead	0.08
Zinc	0.053
Gallium	0.005
Mercury	0.041
Tin	0.031
Cadmium	0.029
Indium	0.029
Aluminium	0.01

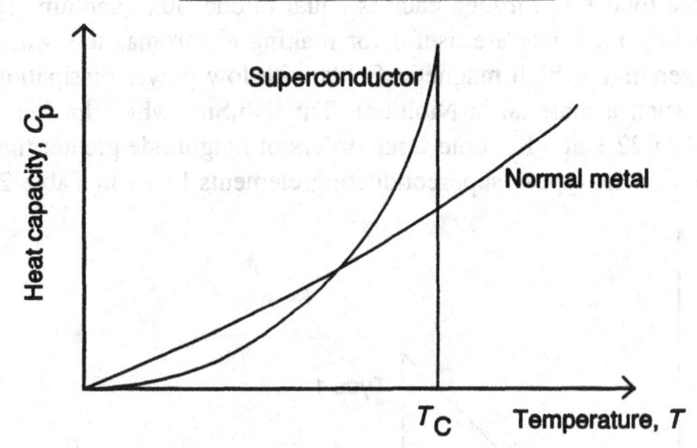

Figure 2.15 *The electronic specific heat capacity at constant pressure,* c_p, *as a function of temperature for a typical superconductor and a normal metal*

entropy is shown in Figure 2.16. In Section 2.2 the low-temperature heat capacity, C, of a normal Fermi gas of electrons was shown to be directly proportional to the temperature. In a superconductor, below its transition temperature, the temperature dependence of C is found to be of the exponential form

$$C = a\exp -\left(\frac{b}{T}\right).$$

The thermal conductivity of superconductors is found to decrease by orders of magnitude as they are cooled below the transition temperature, contrary to what might be expected if, like normal metals, they obeyed the

Wiedemann–Franz law. This behaviour is due to the marked decrease in entropy of the electron system below T_C which results in a corresponding reduction in the amount of heat that can be carried by the electrons.

2.3.2 The theory of superconductivity

It was not until 50 years after the first discovery of superconductivity that a successful microscopic theory was developed. Before that, a number of phenomenological theories were found useful. One of these, the so called two-fluid model of a superconductor, was based on the following postulates; it is worth noting the similarity with the two-fluid model of helium-II described in the next chapter:

1 The electron gas in a superconductor consists of two components. One is devoid of all entropy and is responsible for carrying the supercurrent; it occupies what is called the ground state.

2 The other component carries all of the entropy and behaves as a normal electron gas. It occupies the excited states of the system, which are separated from the ground state by an energy gap.

3 The two components are interpenetrating and non-interacting. The superconducting fraction, f, increases at the expense of the normal fraction $(1-f)$, as the temperature is decreased below T_C. When the conductivity below T_C is measured, the superconducting fraction electrically 'shorts out' the normal fraction.

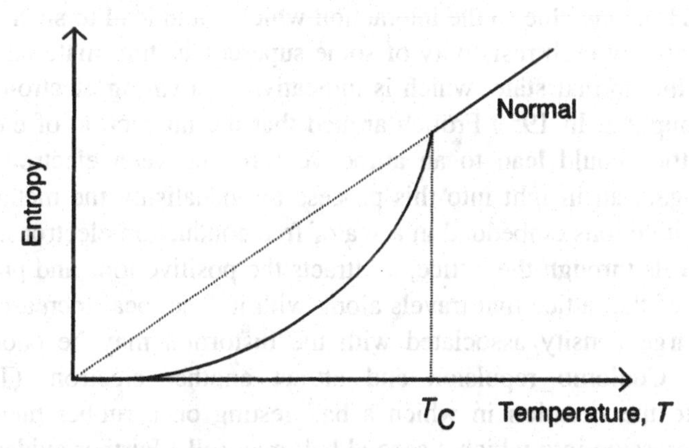

Figure 2.16 *The temperature dependence of the entropy of the electrons in a superconductor*

The model requires no knowledge of the microscopic properties of the ground state or of the mechanism of its formation. Nevertheless, it was successful in giving a semi-quantitative explanation of some of the experimental phenomena. Regarding the specific heat capacity and entropy: When the temperature is reduced, electrons fall into the zero entropy ground state. The temperature where this starts to happen is T_C. The process is complete at 0 K at which all the electrons would end up in the ground state. At a finite temperature, $0<T<T_C$, electrons may be thermally excited across the energy gap. The probability of this happening is proportional to exp (kT/Δ), where Δ is the gap width, thus giving the exponential temperature dependence of the heat capacity. Although the ground state can carry electrical current without dissipation, its zero entropy means that it is unable to carry heat, hence the low thermal conductivity of superconductors. This effect can be put to good use in a superconducting heat switch; a superconducting wire is placed in the bore of an electromagnet, and good thermal contact between the two ends of the wire occurs when the magnet is switched on to drive the wire normal.

The finite resistivity presented to A.C. can also be understood in terms of the two-fluid model: The superconducting ground state has the properties of a pure lossless inductance, this is effectively in parallel with the normal fraction which behaves as a series combination of resistance and inductance. The overall A.C. impedance of this combination has a resistive part which disappears at $T=0$.

The key question is 'What is the origin of the superconducting ground state and the energy gap associated with it?'. It was suspected that it might arise from the pairing of electrons with opposite spin to form a single quasi-particle with properties similar to those of ^{4}He atoms. These could undergo condensation into a kind of 'superfluid' ground state (see Chapter 3 on superfluid ^{4}He). One clue to the interaction which could lead to such pairing is in the relatively high resistivity of some superconducting materials when they are in the normal state, which is indicative of a strong electron lattice (phonon) coupling. In 1950 Frölich argued that the interaction of electrons with the lattice could lead to an attractive force between electrons. It is possible to gain an insight into this process by visualising the metal as an array of positive ions embedded in a sea of free conduction electrons. As an electron travels through the lattice, it attracts the positive ions and produces a distortion of the lattice that travels along with it. The local increase in the positive charge density associated with the distortion may be enough to exceed the Coulomb repulsion and attract another electron. (This is analogous to the situation in which a ball resting on a rubber membrane causes a depression into which a second ball may roll.) Further evidence for the importance of the electron–phonon interaction in superconductors is to be found in the isotope effect. Consider a lattice ion to be a harmonic oscillator with a vibrational energy proportional to $M\omega^2u^2$, where M is the ion mass, $\omega/2\pi$ its vibrational frequency and u its amplitude of displacement.

For a constant energy of interaction, the displacement and lattice distortion is proportional to $M^{½}$. Unfortunately, Frölich was unable to make any further progress with the theory of superconductivity.

The next step was taken seven years later when Cooper showed that the Frölich interaction could lead to the formation of bound pairs of electrons (Cooper pairs) and hence to a reduction in the energy of the electron system as a whole. He also showed that binding electrons with equal and opposite momenta, as well as opposite spin, resulted in the lowest energy for the pair state. Cooper pairs have zero net spin, zero net momentum, a charge of $2e$ and a mass of $2m_e$ (m_e is the mass of a single electron, 9.1×10^{-31} kg). They are central to the BCS theory of superconductivity (Bardeen *et al.*, 1957). The gap, Δ, between the energy of the Cooper pairs and the energy of the single unbound electrons depends on the temperature. At $T \ll T_C$, $\Delta(T) = 1.76kT_C$ and at $T \approx T_C$ it is given approximately by

$$\Delta(T) = 3.1kT_C\left(1 - \frac{T}{T_C}\right)^{1/2}.$$

The single pair binding energy at $T = 0$ is approximately kT_C. The energy gap at $T = 0$ is larger than this owing to cooperative effects whereby the conditions for pair creation are improved if a large number of other Cooper pairs are already in existence. One should not necessarily visualise the Cooper pair as being closely bound in real space. The range of the binding interaction is of the order of the Pippard coherence length, ξ_0, which is related to the transition temperature by

$$\xi_0 \approx 0.2\hbar v_F / kT_C$$

where $v_F = (2E_F/m_e)^{½}$ is the Fermi velocity of the electrons. Normally ξ_0 is about 1 μm so that a Cooper pair can extend over a large number of atomic spacings.

The net momentum of a Cooper pair when carrying a current is just the drift momentum. For a typical superconductor containing about 10^{24} Cooper pairs per m^3 and carrying a current of 10^6 A m^{-2} this only amounts to about $6m_e$ kg m s^{-1}. For comparison, the momentum of an electron travelling at the Fermi velocity, called the Fermi momentum, is of order $10^6 m_e$ kg m s^{-1}. The processes effective in relaxing the momentum of normal electrons have no effect on the net momentum of the Cooper pair. The two electrons comprising the pair may be simultaneously scattered into new states, but as long as the pair is unbroken these must be correlated (i.e. they are of equal and opposite momentum), so that the net momentum of the pair remains unchanged. If the Cooper pairs are broken, then the single electrons may be scattered, as in a normal metal, and a finite conductivity results. This can occur at high current densities corresponding to large drift velocities. The critical velocity depends on the energy gap: Consider a supercurrent flowing in a length of wire of mass m. An equivalent way of picturing this situation is to consider the electrons as being stationary and the wire moving in the opposite direction. Normal scattering processes that transfer energy and or momentum between the wire and the electrons are forbidden as discussed above. However, the wire may slow down through the creation of an excitation within the electron system. This requires a minimum amount of energy, Δ, which can only be provided if the wire is travelling at a sufficiently high speed relative to the electrons. Energy is conserved in the process, therefore

$$\tfrac{1}{2}mv_i - \tfrac{1}{2}mv_f \geq \Delta,$$

where v_i and v_f are respectively the initial and final speeds of the wire. Momentum must also be conserved, and because the excitation is a normal electron with the Fermi momentum, p_F,

$$m(v_i - v_f) = p_F.$$

Eliminating v_f between these two equations gives

$$v_i \geq \frac{\Delta}{p_F} - \frac{p_F}{2m}.$$

Because the wire is massive, the second term on the right hand side is negligible compared with the first and the critical velocity is given by

$$v_c = \frac{\Delta}{p_F}.$$

The current density is $J = nev$, where n is the number of charge carriers per m^3 and e their charge. Therefore the critical current density is given by

$$J_c = \frac{ne\Delta}{p_F}$$

and its value for aluminium is 4.5×10^{11} A m^{-2}. As the temperature approaches T_C the energy gap and hence the critical current tend to zero.

Excitations can also be created optically when the photon energy is comparable to the gap energy. Optical methods are, therefore, one way of measuring the energy gap.

2.3.3 Superconducting tunnel junctions

It is found that an electric current can be passed through an insulating junction between two conductors, provided that the insulating layer is very thin, of the order of a few nanometres. This is possible because of the quantum-mechanical phenomenon of tunnelling: When an electron comes up against one side of the barrier, its quantum-mechanical wavefunction decays with distance into the barrier. However, if the barrier is so thin that the wavefunction does not decay completely to zero by the other side then there is a finite probability that the electron can penetrate the barrier, if energy can be conserved in the process.

A tunnel junction (Figure 2.17a) can be made by sandwiching a thin layer of insulating film or oxide between two conductors. Such a junction between normal metals displays ohmic behaviour because the number of electrons that can tunnel is proportional to the potential difference across the barrier. Figure 2.17b shows the current–voltage characteristics of a tunnel junction formed between two superconductors. At $T = 0$ there is no tunnel current until the voltage across the junction is increased to $2\Delta/e$, whereupon there is a sharp onset of tunnelling. At higher temperatures, though still below T_C, the step in the current becomes less pronounced and occurs at a lower voltage. There is also a residual tunnel current at small voltages.

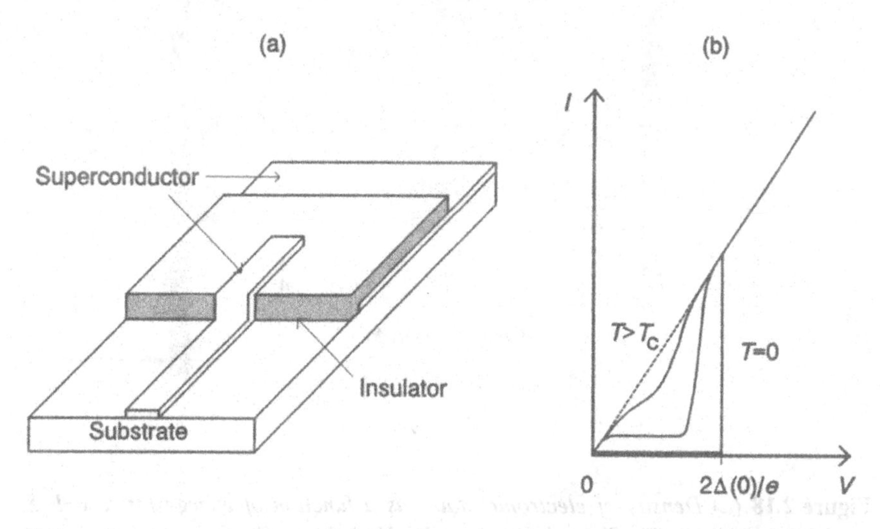

Figure 2.17 *(a) A Superconductor–Insulator–Superconductor (S–I–S) tunnel junction (not to scale). The substrate is a flat insulating medium, such as glass, and the metal is normally deposited by vacuum evaporation. (b) Current–voltage characteristics of an S–I–S tunnel junction; see text for explanation*

The current–voltage characteristics of a tunnel junction can be explained as follows: The probability that an electron is able to tunnel across the barrier and consequently the tunnelling current depends on the density of free-electron quantum states on each side of the barrier. This is because the electron comes from an occupied state on one side of the barrier and tunnels to an empty one on the other. In a superconductor, the density of states is split into two parts. At high energies there is a band of states that are available to single, unpaired, electrons, and at low energies, there are the Cooper pair states. The two bands are separated by the energy gap, Δ. The energy levels of a superconductor–insulator–superconductor (S–I–S) junction, at zero bias, are shown schematically in Figure 2.18a. Although at low temperatures, $T \ll T_C$, most electrons are in the pair states, there are still a few electrons in the single-particle states. If the barrier width is greater than about 2 nm then no tunnelling supercurrent flows. The normal electrons can tunnel to the empty states across the barrier in either direction and the net current is zero. If a small ($V < 2\Delta/e$) voltage is applied across the junction, the energy levels on one side are shifted by an amount Ev with respect to those on the other side. The balance in the single-particle tunnelling between the two sides is upset and a small tunnel current flows.

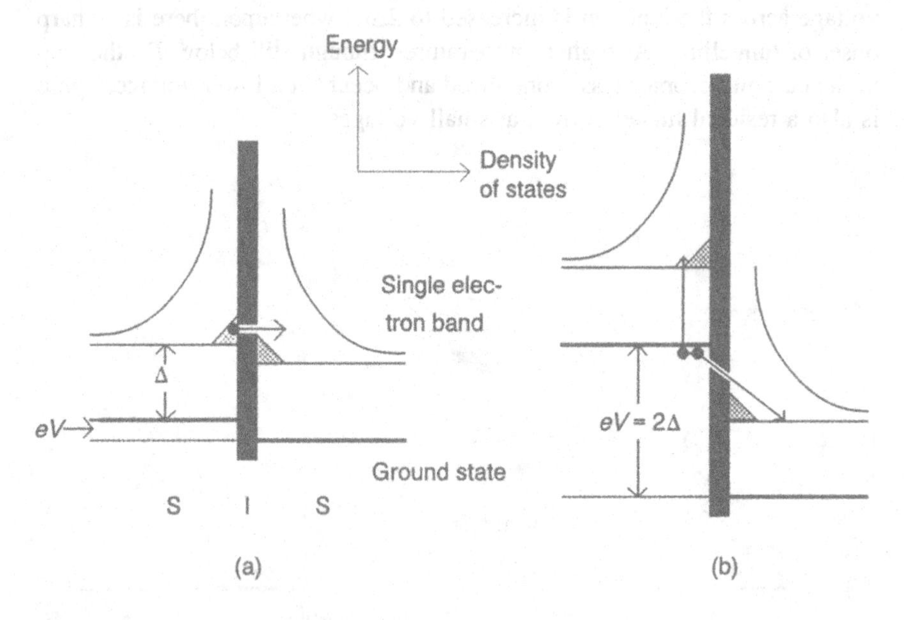

Figure 2.18 *(a) Density of electronic states as a function of energy for a S–I–S tunnel junction at* $0<T\ll T_C$ *and with a small,* $eV<\Delta$*, bias voltage applied. Occupied states are shown cross-hatched. Tunnelling from the thermally populated single electron states is possible. (b) With a much larger bias applied,* $V=2\Delta/e$*, it may become possible for one member of a Cooper pair to penetrate the barrier. This involves supplying an amount of energy equal to* Δ*, since the remaining electron must end up in single particle states, on the same side of the barrier. Therefore, to conserve energy, the tunnelling electron is injected into a level just above the gap, on the other side of the barrier*

The number of single-particle states increases as the temperature is raised and this results in an increased tunnel current at high temperatures. When the voltage is increased to $2\Delta/e$, a second tunnelling process in which energy is conserved becomes possible, shown in Figure 2.18b. A Cooper pair may be split and one electron injected into the single-particle states, on the same side of the barrier. The other electron is free to tunnel across the barrier to the empty single-particle states, just above the gap, on the other side. Because the number of available single particle states is strongly peaked at energies just above the gap, the onset of this process causes a very sudden increase in the tunnel current and provides an accurate means of experimentally determining the energy gap. As the temperature is increased, the gap reduces and the sudden increase in the tunnel current is seen to occur at a lower bias voltage.

If the junction is made thinner than about 2 nm, a tunnelling supercurrent is able to flow. This type of barrier is known as a Josephson junction or weak link. Josephson (1962) showed that in the absence of any potential difference across the junction, the supercurrent density is given by

$$J = J_0 \sin \delta,$$

where δ is the difference in phase between the ground state wavefunctions on each side of the barrier. The critical current density, J_0, depends on the detailed nature of the junction. In the case of a niobium point-contact junction, an example of which is shown in Figure 2.19, J_0 is typically a few

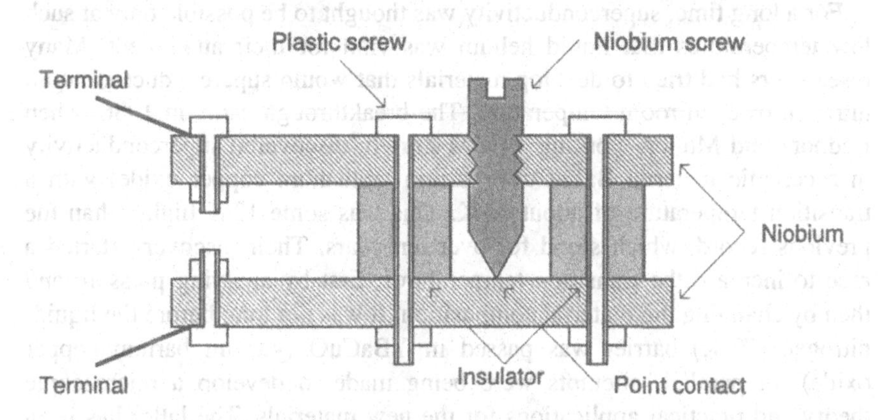

Figure 2.19 *A niobium point contact Josephson Junction*

hundred microamperes at 4.2 K. In the presence of a constant voltage, V, across the junction, the phase difference is time-dependent:

$$\delta(t) = \delta_0 + \frac{2eVt}{\hbar}.$$

The current will, therefore, have a component oscillating at the Josephson frequency of 484 V MHz.

Josephson junctions form the basis of a number of superconducting devices. Small loops of superconductor containing one or two Josephson junctions can be used as highly sensitive magnetometers with a resolution on the scale of a flux quantum. Such devices are known as SQUIDs (superconducting quantum interference devices) and the interested reader is referred to the texts in the bibliography at the end of this chapter for more information about their operation.

2.3.4 Modern developments in superconductivity

The foregoing discussion of superconductivity has touched only the surface of what has proved to be a very rich field of study. Much work on superconductivity has been concerned with applications for conventional superconductors, working at transition temperatures up to about 20 K. For example, there has been steady development of superconducting magnets now capable of producing magnetic fields of up to 20 T, and superconducting devices such as SQUIDs.

For a long time, superconductivity was thought to be possible only at such low temperatures that liquid helium was vital for their attainment. Many researchers had tried to develop materials that would superconduct at liquid nitrogen or even room temperature. The breakthrough came in 1986, when Bednorz and Muller, working at IBM Zurich, discovered superconductivity in a ceramic material BaLaCuO (barium lanthanum copper oxide) with a transition temperature of about 35 K. This was some 12 K higher than the previous record, which stood for over ten years. Their discovery started a race to increase the transition temperatures, first by applying pressure and then by changing the material composition. It was not long before the liquid-nitrogen (77 K) barrier was passed in YBaCuO (yttrium barium copper oxide). In parallel, attempts were being made to develop a microscopic theory and practical applications for the new materials. The latter has been hindered by the mechanical nature of the materials. Ceramics are not readily susceptible to formation into wires which can be wound into high-field solenoids.

Recently reported transition temperatures in the ceramic materials have been in the region of 100 K, somewhat above the ranges of temperature of relevance to this book. However, it is accepted that even high-T_C superconductivity is a low-temperature phenomenon in that quantum-mechanical effects dominate over the thermal disorder. This brings us back to the question considered at the start of this book: Just what is a low temperature?

Bibliography

Bardeen, J., Cooper, L. N. and Schrieffer, J. R. (1957). *Phys. Rev.*, **108**, 1175

Bednorz, G. and Muller, A. (1986). *Z. Phys.*, **B64**, 189

Blakemore, J. S. (1985). *Solid State Physics* (Cambridge: Cambridge University Press)

Deaver, B. S. and Fairbanks, W. M. (1961). *Phys. Rev. Lett.*, **7**, 51

Dekker, A. J. (1970). *Solid State Physics* (London and Basingstoke: Macmillan)

Josephson, B. D. (1962). *Phys. Lett.*, **1**, 251

Mott, N. F. and Davis, E. A. (1979). *Electronic Processes in Non-crystalline Materials*, second edition (Oxford: Clarendon)

Prange, R. E. and Girvin, S. M. (eds.) (1990). *The Quantum Hall Effect* (New York: Springer-Verlag)

Quinn, D. J. and Ittner, W. B. (1962). *J. Appl. Phys.*, **33**, 748

Smith, R. A. (1979). *Semiconductors*, second edition (London: Cambridge University Press)

Ziman, J. M. (1960). *Electrons and Phonons* (Oxford: Clarendon)

General reading on the properties of solids at low temperatures and superconductivity

Bleaney, B. I. and Bleaney, B. (1976). *Electricity and Magnetism*, third edition (Oxford: Oxford University Press)

Feynman, R. P., Leighton, R. B. and Sands, M. (1965). *Lectures on Physics*, vol. III (Reading, Mass: Addison-Wesley)

Kittel, C. (1976). *Introduction to Solid State Physics*, fifth edition (New York: Wiley)

Rose-Innes, A. C. and Rhoderick, E. H. (1977). *Introduction to Superconductivity.* (London: Pergamon)

3

Properties of liquid helium

At saturated vapour pressure, the two isotopes of Helium, ^{3}He and ^{4}He, remain liquid down to the lowest temperatures. They are the so-called quantum fluids and exhibit a wide range of interesting phenomena, some of which will be reviewed in this chapter. These liquids are also widely used in the attainment of low temperatures down to a few millikelvins. The reason why helium remains liquid at such low temperatures is that the zero point or ground state motion of the helium atoms is able to outbalance the weak van der Waals forces attracting the atoms together. The two isotopes behave similarly near their boiling points at atmospheric pressure (Table 3.1); they are like dense classical gases. However, as the temperature is reduced, there is a marked divergence in their respective properties. This is because ^{4}He atoms are bosons, having integral spin and a symmetric wavefunction, any number of them being allowed to occupy a particular quantum state, whereas, like electrons, ^{3}He atoms are fermions and so obey the Pauli exclusion principle. The low-temperature properties of each liquid will be considered separately in Sections 3.1. and 3.2.

3.1 Liquid ^{4}He

The phase diagram of ^{4}He is shown in Figure 3.1. The λ-line divides the region where the helium behaves as a dense classical gas, known as helium-I, from the low-temperature region where its properties are very different and it is termed helium-II. The specific heat, for instance, was observed to exhibit a discontinuous change in its behaviour as the temperature or pressure is changed through the λ-line (Figure 3.2). The sudden changes in the properties of ^{4}He are very obvious when a bath of the liquid is observed as it is cooled through the λ-transition; it is seen to stop boiling abruptly (although it still vaporises from the surface). Further experimental measurements on helium-II showed it to possess even more interesting properties: Its viscosity, for instance, was found to be very low and dependent on the method used for its measurement (Figure 3.3). In one technique, the flow of liquid through a very fine capillary was measured, the

Table 3.1 Properties of liquid ^3He and ^4He near their normal boiling points

Property	^3He	^4He
Boiling point at 1 atm	3.2 K	4.22 K
Critical temperature	3.32 K	5.2 K
Density, kg m^{-3}	60	130
Latent heat, J kg^{-1}	8000	20 000

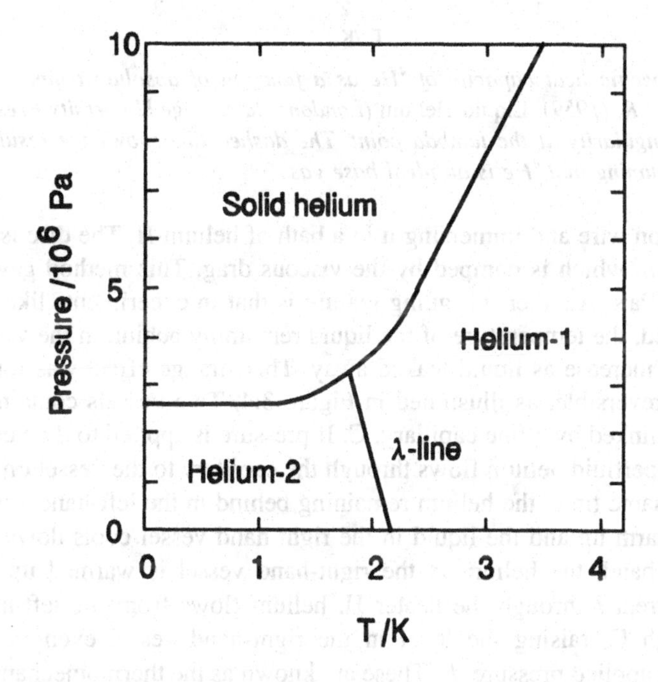

Figure 3.1 *P–T phase diagram of ^4He*

rate of flow being related to the viscosity by Poiseuille's equation. By this method, values of the viscosity obtained were of the order 10^{-12} Pa s (for comparison at room temperature water has a viscosity of approximately 10^{-4} Pa s). A second method of measurement involved suspending a disc at the

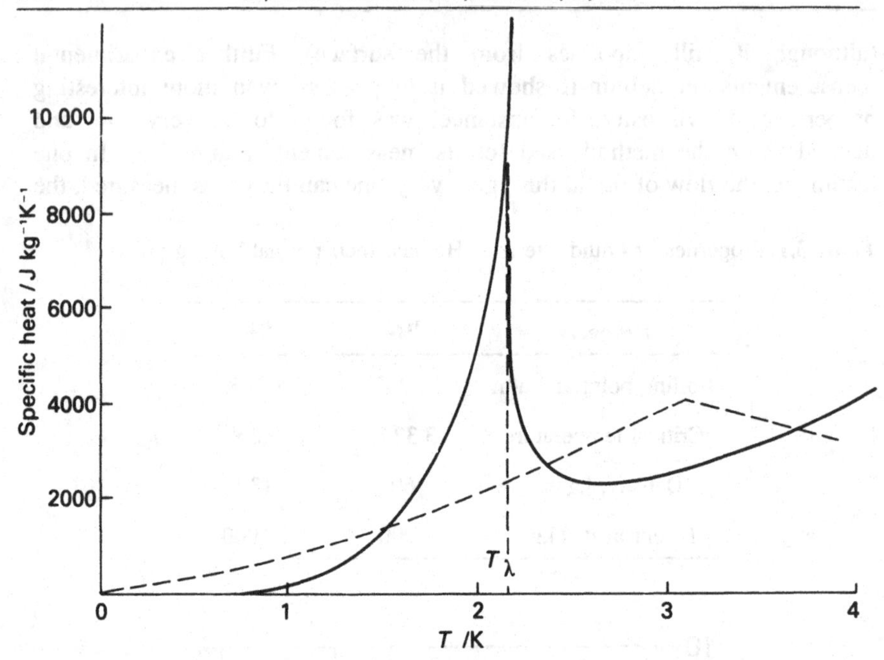

Figure 3.2 *Specific heat capacity of ^4He as a function of absolute temperature [after Atkins, K. R. (1959).* Liquid Helium *(London: Cambridge University Press)], showing the singularity at the lambda point. The dashed line shows the result of calculation assuming that ^4He is an ideal bose gas*

end of a torsion wire and immersing it in a bath of helium-II. The disc is set into oscillation, which is damped by the viscous drag. This method gave a value of 10^{-6} Pa s. Another intriguing feature is that in experiments like the first mentioned, the temperature of the liquid remaining behind in the vessel was found to increase as liquid leaked away. This strange effect was found to be totally reversible, as illustrated in Figure 3.4. Two vessels containing helium-II are linked by a fine capillary, C. If pressure is applied to the vessel on the left, superfluid helium flows through the capillary to the vessel on the right. At the same time, the helium remaining behind in the left-hand vessel is found to warm up and the liquid in the right-hand vessel cools down. If, on the other hand, the helium in the right-hand vessel is warmed up, by passing a current I through the heater H, helium flows from the left-hand vessel through C, raising the level in the right-hand vessel even in the absence of an applied pressure, P. These are known as the thermomechanical effects. Finally, if helium-II is in contact with a solid surface, there is a tendency for a film to form on the surface a few atoms thick. This is not peculiar to helium, but helium-II is unusual in that the film flows rapidly over the surface and the liquid can actually leak out of an open-topped container! The thickness, d, of the film on a vertical surface is due to the balance between the van der Waals forces attracting the helium to the surface and gravity pulling it back down, in fact

$$d = \frac{A}{h^{1/3}},$$

and it amounts to about 20 nm at $h \approx 10$ mm.

The velocity of the film flow amounts to a few centimetres per second.

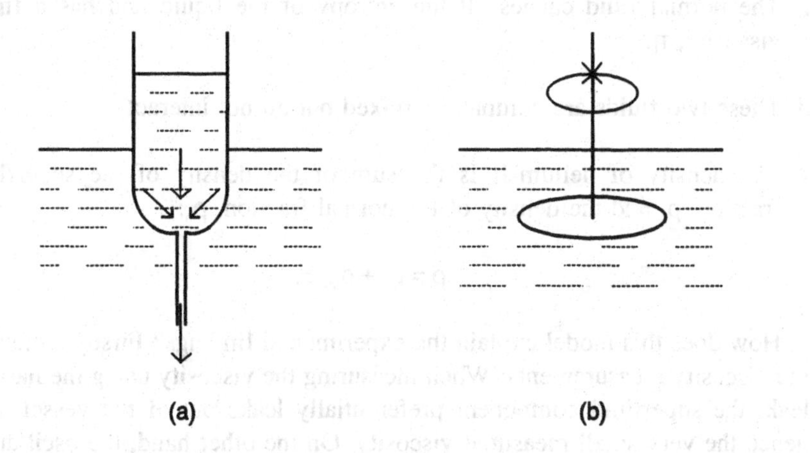

Figure 3.3 *Schematic diagram showing two methods of measuring the viscosity of helium-II, (a) by measuring the rate of leak through a narrow capillary and (b) by observing the damping of an oscillating disc*

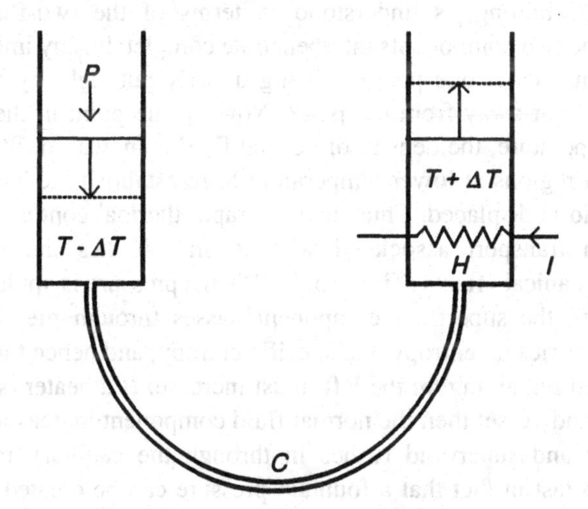

Figure 3.4 *Schematic diagram of an experiment to illustrate the thermomechanical effects in helium-II. The two reservoirs, left and right, are linked by a narrow capillary, C, which allows only the superfluid to pass through*

In order to explain these phenomena, the two-fluid model for helium-II was proposed, based on the following postulates:

1 Helium-II behaves as if it was a mixture of two fluids, one quite normal like helium-I, the other being completely devoid of viscosity and entropy and known as a superfluid.

2 The normal fluid carries all the entropy of the liquid and has a finite viscosity, η.

3 These two fluids are completely mixed but do not interact.

4 The density of helium-II is the sum of the density of the superfluid fraction, ρ_s and the density of the normal fraction, ρ_n:

$$\rho = \rho_n + \rho_s.$$

How does this model explain the experimental findings? Firstly, consider the viscosity measurements: When measuring the viscosity using the narrow leak, the superfluid component preferentially leaks out of the vessel, and hence the very small measured viscosity. On the other hand, the oscillating disc is damped only by the normal component, leading to the higher value for the viscosity.

The very high thermal conductivity of helium-II, which leads to the cessation of boiling, is understood in terms of the two-fluid model as follows: The two components interpenetrate completely, any imbalance in the relative densities at one position being quickly put right by a flow of the superfluid to or away from the point. Now, if one point in the fluid is at a higher temperature, the density of normal fluid is increased. Superfluid then flows from regions at lower temperature to re-establish the balance and the normal fluid is displaced. Thus, there is rapid thermal conduction and a net momentum transport associated with it. In this lies the origin of the thermomechanical effects (Figure 3.4). When a pressure is applied to the left-hand vessel, the superfluid component passes through the fine capillary; because it carries no entropy, the specific entropy, and hence the temperature of the liquid remaining on the left, must increase. If a heater is turned on in the right-hand vessel then the normal fluid component increases markedly in its locality and superfluid rushes in through the capillary to redress the balance, so fast in fact that a fountain pressure can be created (Figure 3.5).

Liquids can carry sound waves and helium-II is no exception, but there are new modes of propagation which arise from the two-fluid properties. First sound, as it is known, is due to density oscillations of the whole fluid while the entropy remains constant, i.e. there is no relative movement of the normal and superfluid components. First sound can be excited in the conventional manner by means of a mechanical transducer in the liquid. However, it is also possible for temperature waves to propagate in helium-II, that is, oscillations in entropy at constant density; this implies relative oscillatory movement of the normal and superfluid fractions and is known as second sound. It may be excited by pulsing a heater in the liquid at high frequency. Third sound and fourth sound are modes transmitted through the superfluid film and narrow capillaries respectively (McClintock *et al.*, 1984).

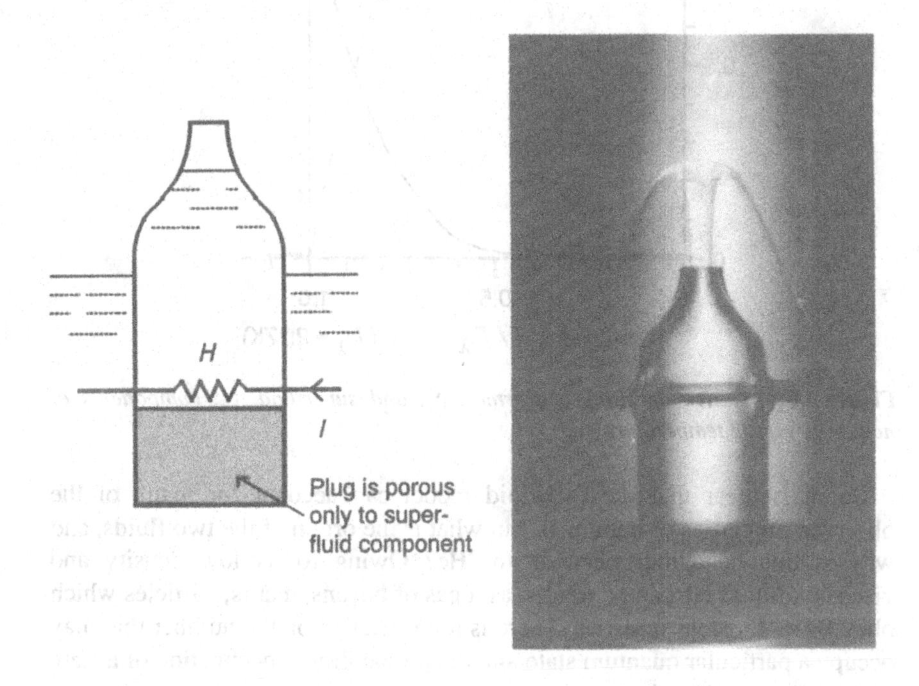

Figure 3.5 *(Left) Experimental arrangement for demonstrating the fountain effect with helium-II. A fountain pressure is produced by supplying a current to the heater, H, which raises the temperature of the helium inside the vessel, resulting in a rapid inflow of superfluid through the plug at the bottom. (Right) Photograph of the fountain effect (courtesy of J. G. M. Armitage, Department of Physics, University of St. Andrews)*

Measurements of the relative densities of the superfluid and normal components and their dependence on temperature (Figure 3.6) were made by Andronikashvili in 1946. He immersed a torsion pendulum consisting of a stack of closely spaced discs in helium-II. The spacing was so close (≈ 0.2 mm) that the normal component was trapped between the discs and rotated with them, while the superfluid component was stationary. The inertia of the pendulum and hence its period was, therefore, dependent on the ratio of the normal to the superfluid components. It is interesting to note that it is only necessary to cool ^{4}He to 1 K for it to be almost totally superfluid.

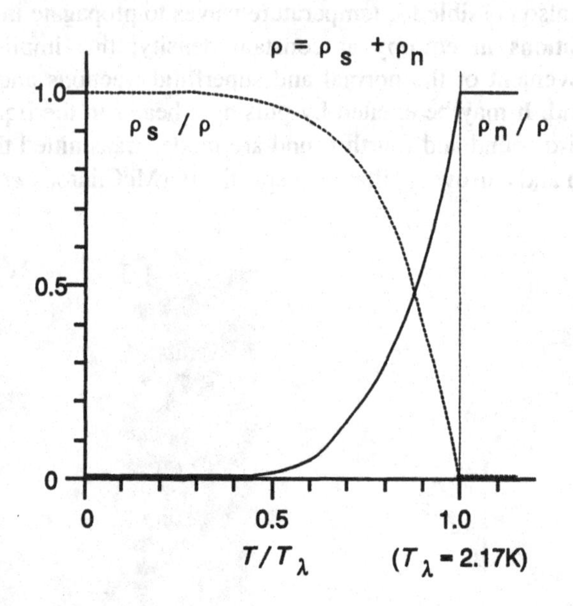

Figure 3.6 *Relative densities of normal,* ρ_n*, and superfluid,* ρ_s*, components of helium-II versus temperature*

So, it is clear that the two-fluid model can account for many of the observed properties of helium-II, but what is the origin of the two fluids, and why is this behaviour peculiar to ^{4}He? Owing to its low density and viscosity, liquid He can be treated as a gas of bosons, that is, particles which obey Bose–Einstein statistics. There is no restriction on the number that may occupy a particular quantum state and the probability of occupation of a state of energy E is given by

$$P(E) = \left\{ \exp\left(\frac{E - \mu}{kT} \right) - 1 \right\}^{-1},$$

where μ is the chemical potential. If a gas of bosons is cooled to very low temperatures then it is possible for an appreciable number of atoms to 'condense' into the ground state, a phenomenon known as Bose–Einstein condensation. In 1938 London suggested that the λ-transition in ^{4}He may be a manifestation of Bose condensation. The reasoning behind this is as follows: The quantum energy levels of a particle moving in a box of volume V are given by

$$E = \frac{h^2}{2mV^{3/2}}(a^2 + b^2 + c^2),$$

where a, b and c are integers (see for instance any basic text on quantum mechanics). If there is a large number of particles in the box, this result still holds for each one, as long as there are no interactions. When the volume of the system is large, the states are very close together in energy. Under these conditions, it is best to describe the variation with energy of the number of discrete states in terms of the density of states function, $D(E)$. The number of allowed energy levels in the narrow band between E and $E + \mathrm{d}E$ is given by $D(E)\,\mathrm{d}E$, where

$$D(E) = 2\pi V(2m)^{3/2}\frac{E^{1/2}}{h^3}.$$

The total number of atoms in the box, N_e, is obtained by multiplying the number of available states of energy E by the probability of occupation of each state, and then summing over all energies.

$$N_e = \int_0^\infty D(E)P(E)\,\mathrm{d}E = \frac{2\pi V(2m)^{3/2}}{h^3}\int_0^\infty \frac{E^{1/2}\mathrm{d}E}{\exp\left(\dfrac{E-\mu}{kT}\right) - 1}.$$

The chemical potential, μ, must be negative, because if it were not so then it would be possible to have the bizarre situation where there would exist a negative probability of occupation of states with energy less than μ; therefore, $\exp(E/kT) > \exp(E\text{-}\mu/kT)$ and

$$N_e \le \frac{2\pi V(2m)^{3/2}}{h^3}\int_0^\infty \frac{E^{1/2}\mathrm{d}E}{\exp(E/kT) - 1}.$$

Evaluating the integral, either numerically or using tables, gives

$$N_e \le \frac{41.1V(mkT)^{3/2}}{h^3}.$$

It is clear that as T approaches zero so does N_e, so where do the atoms go? The answer lies in the fact that so far only the excited states of the system have been considered, while the ground state for which $a = b = c = 0$ has been ignored. The total number of particles must be conserved, so as the temperature is reduced below the point at which the excited states can accommodate them all, they must be forced into the ground state (Figure 3.7). Assuming that this starts to happen at some temperature T_B, the Bose condensation temperature, London showed that for $T \approx T_B$, μ is very small and can be taken to be zero, hence a value for T_B is obtained,

$$T_B = \frac{h^2}{11.9mk}\left(\frac{N}{V}\right)^{2/3},$$

where N is the total number of Bosons in the ground plus excited states,

$$N = N_0 + N_e.$$

Using the appropriate data for ^{4}He gives $T_B = 3.1$ K, certainly close to 2.17 K, the difference being due to interactions between the atoms in helium (it is after all a liquid). The ratio of the number of particles in the ground state to the total number in the system as a whole is given by

$$\frac{N_0}{N} = 1 - \left(\frac{T}{T_B}\right)^{3/2}.$$

This falls to zero at $T = T_B$ as expected and tends to 1 as T approaches absolute zero.

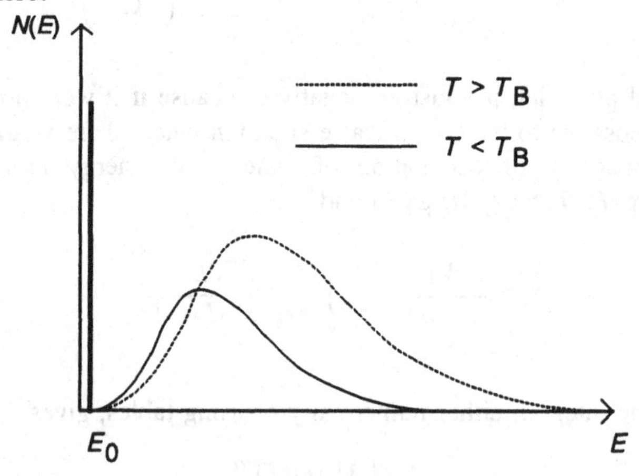

Figure 3.7 *Distribution of Bosons among energy states in an ideal Bose gas. When the gas is cooled to below the Bose condensation temperature the ground state suddenly becomes occupied*

The specific heat is due only to the particles in excited states, to each of which can be apportioned an average thermal energy $3kT/2$; therefore,

$$c \approx \frac{\mathrm{d}}{\mathrm{d}T}\left\{\frac{3NkT^{5/2}}{2T_B^{3/2}}\right\} = \frac{15R}{4}\left(\frac{T}{T_B}\right)^{3/2} \mathrm{J\,K^{-1}\,mol^{-1}}.$$

This result is compared with the experimental data in Figure 3.2.

So, there is a strong weight of evidence for the superfluid being the Bose condensate of ^{4}He. This isn't the whole story, though; at very low temperatures ^{4}He is almost totally superfluid in nature. However, quantities like the specific heat do not vary exactly as predicted by the simple theory above. At temperatures below 0.6 K, the specific heat varies as T^3. This suggests the presence of phonons in the superfluid; in fact, these are just one of the types of excitation due to collective modes of the system at low temperatures. But how is it that these excitations do not break down the superfluid state? It is due to the peculiar dispersion curve or energy–momentum relationship of the excitations that may be deduced from neutron-scattering experiments (Figure 3.8). At low energies there are phonons with a velocity equal to that of first sound; the peculiar minimum at higher energies is attributed to excitations called rotons, the precise nature of which little is known. Now, consider superfluid helium flowing through a capillary tube. In the frame of reference of the helium, the superfluid is stationary and the capillary tube is moving in the opposite direction. In order to slow down, the tube must lose energy which is converted into excitations within the helium. This requires the tube to be travelling faster than a certain critical velocity, v_c, known as the Landau critical velocity. Both energy and momentum must be conserved in the creation of excitations; therefore, by analogy with the calculation of the critical current of a superconductor (see Section 2.3.2), v_c is given by

$$v_c = \left\langle\frac{\varepsilon(p)}{p}\right\rangle_{\mathrm{min}},$$

where $\varepsilon(p)$ is the energy and p the momentum of the excitation. The minimum of $\varepsilon(p)/p$ is close to the roton minimum on the dispersion curve, therefore the tube loses energy by the excitation of roton modes. The value of the critical velocity turns out to be in the region of 50 m s^{-1}.

To complete this discussion of the basic properties of ^{4}He, consider what happens if the vessel containing the superfluid is spun. At first, it might be expected that the superfluid will remain stationary (there is no viscous drag to transmit the rotation to the fluid) and so the surface of the fluid would remain level. This follows from the mathematical statement of superfluidity in the two-fluid model,

$$\text{curl } v_s = 0,$$

where v_s is the velocity of the superfluid component. Experimentally, however, the fluid surface takes a concave form, which means that it must be rotating. London and Onsanger suggested that helium-II is threaded by a set of quantised vortex lines (Figure 3.9) which carry the angular momentum associated with the rotational motion, but when the whole volume is considered the above equation is still satisfied. The angular momentum of circulation of the vortex lines is quantised in units of $\hbar/m$ (Vinen, 1961). They can be visualised as the rotation of the liquid around a 'hole' at $r = 0$ with a velocity profile $v = A/r$, where $A = n\hbar/m$ and n is an integer. The radius of the hole is determined by the balance of centrifugal force and surface tension; it turns out to be 0.03 nm which is less than the atomic dimension (~ 0.1 nm). So, in reality, there is no macroscopic hole in the liquid.

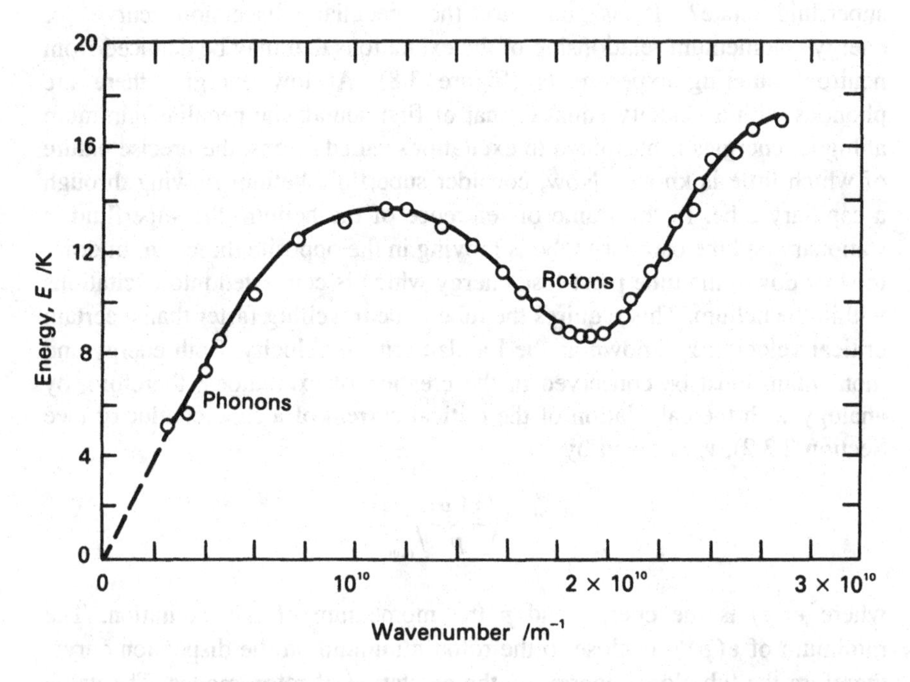

Figure 3.8 *Dispersion curve of excitations in helium-II, as deduced from neutron scattering measurements [after Henshaw, D. G. and Woods, A. D. B. (1961)* Phys. Rev., *121, 1266]*

3.2 Liquid ^{3}He

Upon cooling to below 1 K, ^{3}He undergoes a gradual change in its properties, which finally becomes complete at temperatures near 50 mK. Its specific heat, for example, which varies with temperature in the same way as that of helium-I near to the boiling point, shows an approximately linear dependence on T below 50 mK (Figure 3.10). Changes are also seen in the temperature dependence of its viscosity (Figure 3.11), thermal conductivity (Figure 3.12) and magnetic susceptibility (Figure 3.13). The last of these properties is particular to the lighter isotope; its magnetism arises because of the intrinsic quantised angular momentum, or spin, of the nucleus, which is also the origin of its completely different nature to ^{4}He. The ^{3}He atom has a net spin ½ and so an antisymmetric wavefunction; like electrons, it obeys the Pauli exclusion principle and hence Fermi–Dirac statistics. The change in the properties of ^{3}He is associated with the transition to a degenerate Fermi gas at $T < T_F$, where T_F is the Fermi temperature E_F/k.

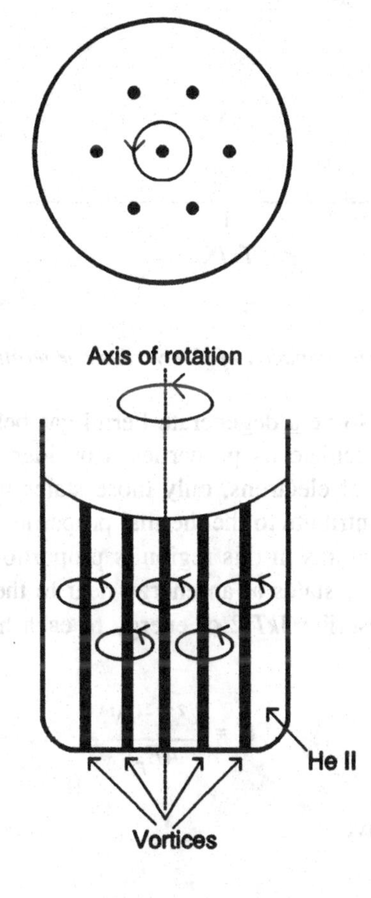

Figure 3.9 *Appearance of quantised vortices in a spinning bucket of helium-II*

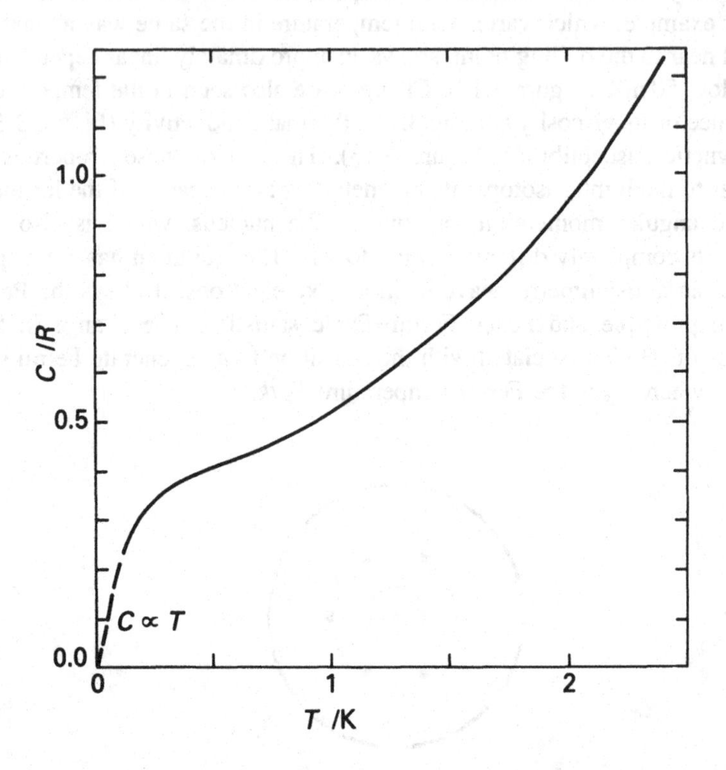

Figure 3.10 *Specific heat capacity of* ^3He *versus temperature [after Wilks (1970)]*

If we assume ^3He to be a degenerate Fermi gas below 50 mK, it is fairly straightforward to calculate its properties. Consider first the specific heat: Just like in the case of electrons, only those states within kT of the Fermi energy are able to contribute to the thermal properties of a degenerate Fermi gas. The number of atoms in this region is proportional to $kT \cdot D(E_F)$ where $D(E_F)$ is the density of states at an energy near to the Fermi energy, $D(E_F) = 3N/2E_F$, so if we ascribe $3kT/2$ of energy to each helium atom we obtain a thermal energy,

$$U = \frac{k^2 T^2 9N}{4E_F} .$$

Exact calculations give

$$U = \frac{k^2 T^2 \pi^2 N}{4E_F}.$$

The specific heat is just $dU/dT = k^2 T\pi^2 N/2E_F$, which is proportional to T as found experimentally. At higher temperatures, $kT \sim E_F$; E_F is found to depend on temperature, so this simple argument breaks down. The viscosity of a gas is given, according to simple kinetic theory arguments, by

$$\eta = \frac{1}{3}\rho\tau\bar{v}^2,$$

where ρ is the density, $\bar{v}$ the mean molecular velocity and τ the mean scattering time. In the case of a degenerate Fermi system $\bar{v}$ is replaced by the Fermi velocity, v_F, and is independent of temperature. The scattering rate τ^{-1} is proportional to the probability that two atoms, initially in states E_1 and E_2, collide and scatter into new states E_3 and E_4; therefore, the scattering rate depends on the probability of finding *two* empty states within kT of E_F. This is proportional to T^2, and hence

$$\eta \propto T^{-2},$$

in agreement with experimental measurements. The thermal conductivity is also given by simple kinetic theory, in terms of the viscosity

$$K = \eta c,$$

where c is the specific heat, $c \propto T$; therefore

$$K \propto T^{-1},$$

also in agreement with experiment.

As a consequence of its nuclear spin, a ^3He atom has a small magnetic dipole moment. In the absence of an applied magnetic field, the dipoles are randomly oriented, owing to thermal motions. When a field is applied, they line up, giving rise to a small paramagnetic susceptibility, χ. Classically, the susceptibility depends on the ratio of the magnetic to thermal energies, $\mu_N B/kT$, and so is proportional to $1/T$. This is Curie's law and is obeyed by ^3He at high temperatures; it also applies to many other magnetic materials (see Section 6.1). However, as ^3He is cooled, the susceptibility becomes independent of temperature. This is seen to arise as follows: Without a magnetic field applied, the energy levels corresponding to the two possible values of nuclear spin quantum number are the same, and they are equally occupied. When the field is switched on, the energy levels corresponding to

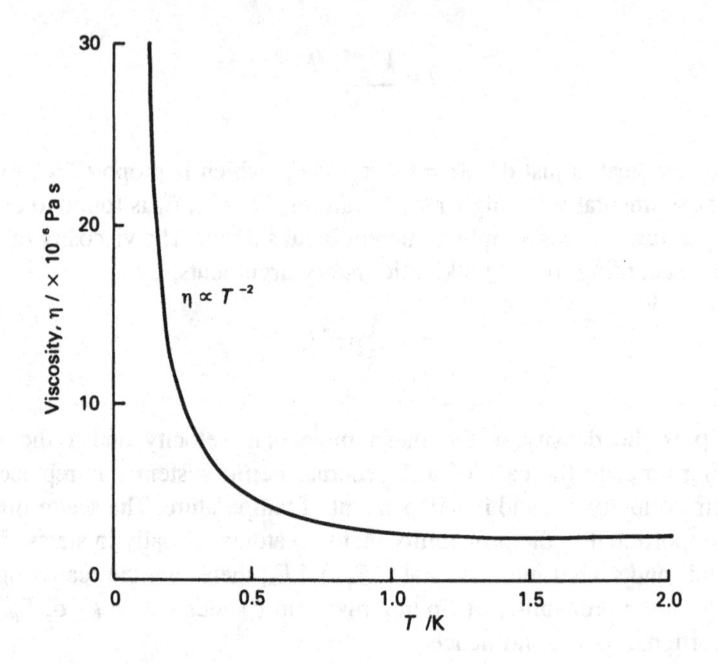

Figure 3.11 *Viscosity of ^{3}He versus temperature [after Wilks (1970)]*

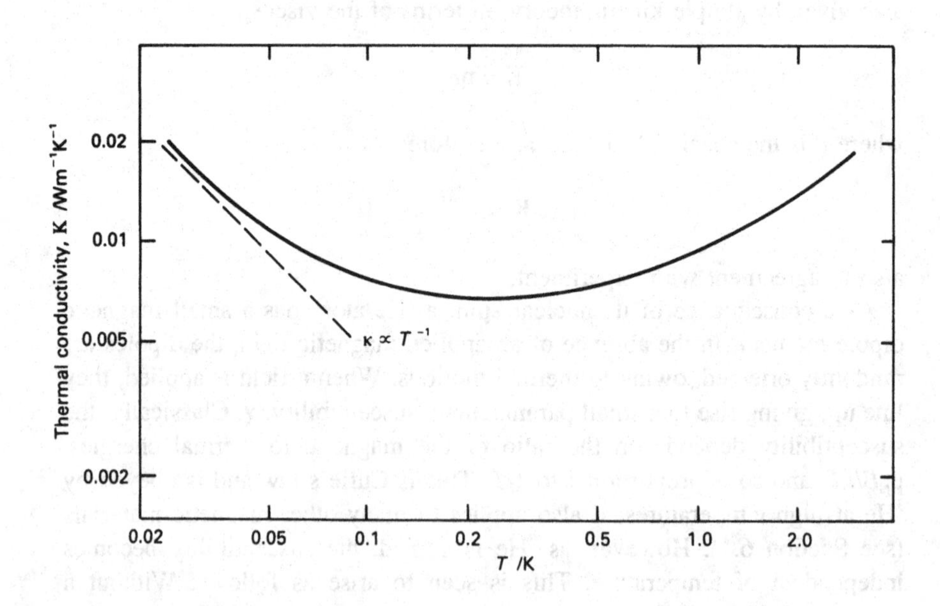

Figure 3.12 *Thermal conductivity of ^{3}He versus temperature*

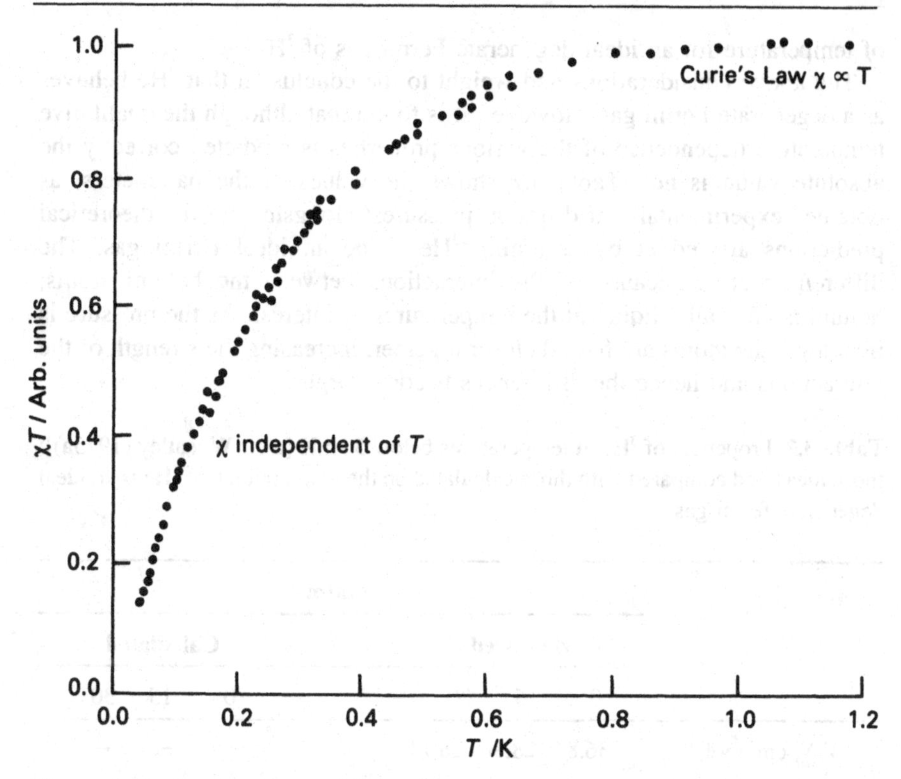

Figure 3.13 *Magnetic susceptibility of ^{3}He versus temperature [after Thomson, A. L., Meyer, H. and Adams, E. D. (1962)* Phys. Rev., ***128****, 509]*

spins lining up parallel to the field are shifted down in energy by $\mu_N B$ (μ_N is the nuclear magneton; $\mu_N = 5.051 \times 10^{-27}$ JT^{-1}). The levels corresponding to antiparallel spins are shifted up in energy by the same amount. The ^{3}He atoms now redistribute themselves among the levels so that the Fermi energy remains unchanged. If the magnetic energy, $\mu_N B$, is much smaller than the Fermi energy, E_F, which is normally the case except at very high field strengths, the number of spins transferring from the antiparallel to the parallel states, ΔN, is equal to the number of occupied states that were shifted to energies above the Fermi level when the field was switched on. At low temperatures, such that $kT \ll E_F$, $\Delta N = D(E_F)\,\mu_N B$, where $D(E_F)$ is the density of states at the Fermi energy. This results in an excess number, $2\Delta N$, of spins lining up parallel to the applied field and, therefore, a net magnetic moment per unit volume,

$$\boldsymbol{M} = 2\mu_N \Delta N = \frac{2D(E_F)\,\mu_N^2 \boldsymbol{B}}{V} = \frac{2m}{2\pi^2\hbar^2}\left(\frac{3\pi^2 N}{V}\right)^{1/3} \mu_N^2 \boldsymbol{B}.$$

The paramagnetic susceptibility per unit volume is given by $\chi = \mu_0 \boldsymbol{M}/\boldsymbol{B}$, where μ_0 is the magnetic permeability of free space, χ is clearly independent of temperature for an ideal degenerate Fermi gas of ^{3}He.

The above considerations add weight to the conclusion that ^{3}He behaves as a degenerate Fermi gas. However, it is found that although the qualitative temperature dependence of the various properties is predicted correctly the absolute value is not. Table 3.2 shows the values of the parameters, as obtained experimentally at different pressures; alongside are the theoretical predictions arrived at by assuming ^{3}He to be an ideal Fermi gas. The differences arise because of the interactions between the helium atoms; helium is after all a liquid at the temperatures of interest. As the pressure is increased, the atoms are forced closer together, increasing the strength of the interactions and hence the differences become larger.

Table 3.2 Properties of ^{3}He at temperatures below 50 mK [after Wheatley (1975a)]: those measured compared with those calculated on the assumption that ^{3}He is an ideal degenerate fermi gas.

	P/atm					
	Measured			Calculated		
	0	15	30	0	15	30
V/N, cm^3 mol^{-1}	36.84	28.86	26.14	–	–	–
C/RT, mol^{-1} K^{-1}	3.0	4.0	4.6	1.0	0.85	0.8
ηT^2, Pa s	0.18	0.14	0.10	0.56	0.65	0.58
$\chi/10^{-8}$	10.6	21.5	28.9	1.16	1.26	1.30

A successful theory due to Landau (1957) considers the motion of the helium atoms in the interaction field of all their neighbours. He conceptually replaces the helium atoms by corresponding quasi-particles which obey Fermi statistics. The quasi-particles have an effective mass m^* that is higher than the bare ^{3}He mass, and the first effect of this is to reduce the Fermi energy relative to an ideal gas of ^{3}He. Simply defining an effective mass is sufficient to explain the difference in the specific heat,

$$c = \left(\frac{m^*}{m}\right) c_{ideal},$$

but not the other properties. The success of the Landau theory is that it is able to account for all of the properties in terms of a set of phenomenological parameters, the temperature dependencies remaining the same as for an ideal Fermi gas, but the absolute values coming into line with experiment.

At still lower temperatures, 1–3 mK, ^{3}He is observed to undergo two phase transitions (Figure 3.14). The two phases, A and B, were observed to have some properties in common with helium-II and have, therefore, been called the superfluid phases of ^{3}He. The details of the phase diagram change when a magnetic field is applied to the system. In zero field, there exists a point known as the polycritical point (PCP), where the A- and B-phases exist in equilibrium, whereas in a magnetic field, the A-phase exists to zero temperature and a new phase denoted A1 separates it from the normal phase above the transition temperature, T_C.

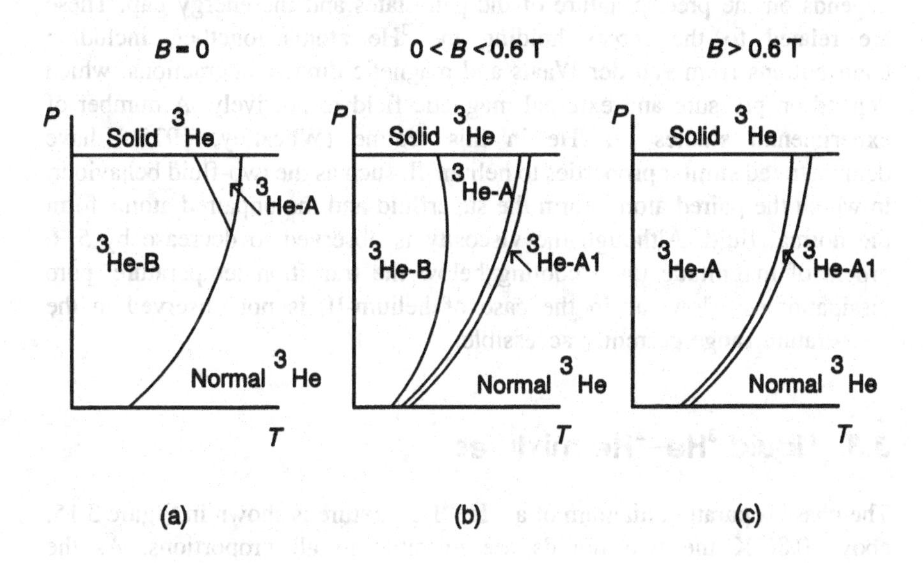

Figure 3.14 *Schematic representation of low temperature phase diagram for ^{3}He showing the superfluid phase transitions, (a) in zero magnetic field, (b) and (c) in a finite applied field. The polycritical point PCP is at a temperature of about 2.4 mK and a pressure of 2.2 MPa*

It had been theoretically predicted as early as 1959 that a superfluid state of ^{3}He might exist by applying the proven BCS theory of superconductors (see Chapter 2). There followed a great deal of abortive experimental search for the transition. At the same time, the theoretically predicted transition temperatures were getting further and further out of the experimentalists' reach. The first indication of a transition came as late as 1971; the discovery was made by Osheroff at Cornell University, using low temperatures and high pressures in a Pomerantchuk cell; see Section 5.3. The high pressure seems necessary in order to increase the interactions that lead to the pairing of the ^{3}He atoms to form the new phases. The rapid development in low-temperature technology and the application of nuclear cooling (see Chapter 6) enabled the transition to be observed at zero pressure and a temperature of 1 mK.

In the superfluid phases, the ^{3}He atoms form correlated pairs with equal spin ($S = 1$). The Pauli exclusion principle then requires that the angular momentum of the pair is unity ($L = 1$), known as p-wave pairing. One can imagine the two ^{3}He atoms orbiting around each other, though their separation in space is as much as 200 atomic dimensions at zero pressure. There is an energy gap, ΔE, between the pair states and the continuum of single particle states at higher energy, hence the pairs are stable at temperatures, $kT \ll \Delta E$. The distinction between the A-, A1- and B-phases depends on the precise nature of the pair states and the energy gap. These are related to the forces holding the ^{3}He atoms together, including contributions from van der Waals and magnetic dipolar interactions, which depend on pressure and external magnetic field respectively. A number of experimental studies on ^{3}He in this regime (Wheatley, 1975b) have demonstrated similar properties to helium-II, such as the two-fluid behaviour, in which the paired atoms form the superfluid and the unpaired atoms form the normal fluid. Although the viscosity is observed to decrease by 5–6 orders of magnitude upon cooling below the transition temperature, pure dissipationless flow, as in the case of helium-II, is not observed in the temperature range currently accessible.

3.3 Liquid ^{3}He—^{4}He mixtures

The phase separation diagram of a ^{3}He—^{4}He mixture is shown in Figure 3.15; above 0.86 K the two liquids are miscible in all proportions. As the temperature is lowered, the mixture can become superfluid, the value of the transition temperature being dependent on the ^{3}He concentration. At temperatures less than 0.86 K, phase separation takes place, with the ^{3}He-rich (concentrated) phase floating on top of the more dense ^{4}He-rich (dilute) phase. The mixture ratio in the two phases at any temperature is given by the points where a horizontal line corresponding to the temperature of interest crosses the phase separation curve, C and D in the diagram. Of particular interest is that, at very low temperatures, there is a finite solubility of ^{3}He in ^{4}He of approximately 6.5 per cent. This implies that for concentrations up to 6.5 per cent, less energy is required to add an atom of ^{3}He to superfluid ^{4}He than to add it to pure ^{3}He.

This system is of great interest, as low-density solutions of ^{3}He in ^{4}He behave as an ideal Fermi gases of variable concentration. At temperatures less than about 0.5 K, few phonons or rotons are excited in the ^{4}He. The superfluid carries no entropy and has little effect on the properties of the mixture; it acts as a kind of 'mechanical vacuum' for the ^{3}He to move in. Between 100 mK and 500 mK the mixture has the properties of a dilute classical gas of ^{3}He, for example its specific heat is independent of

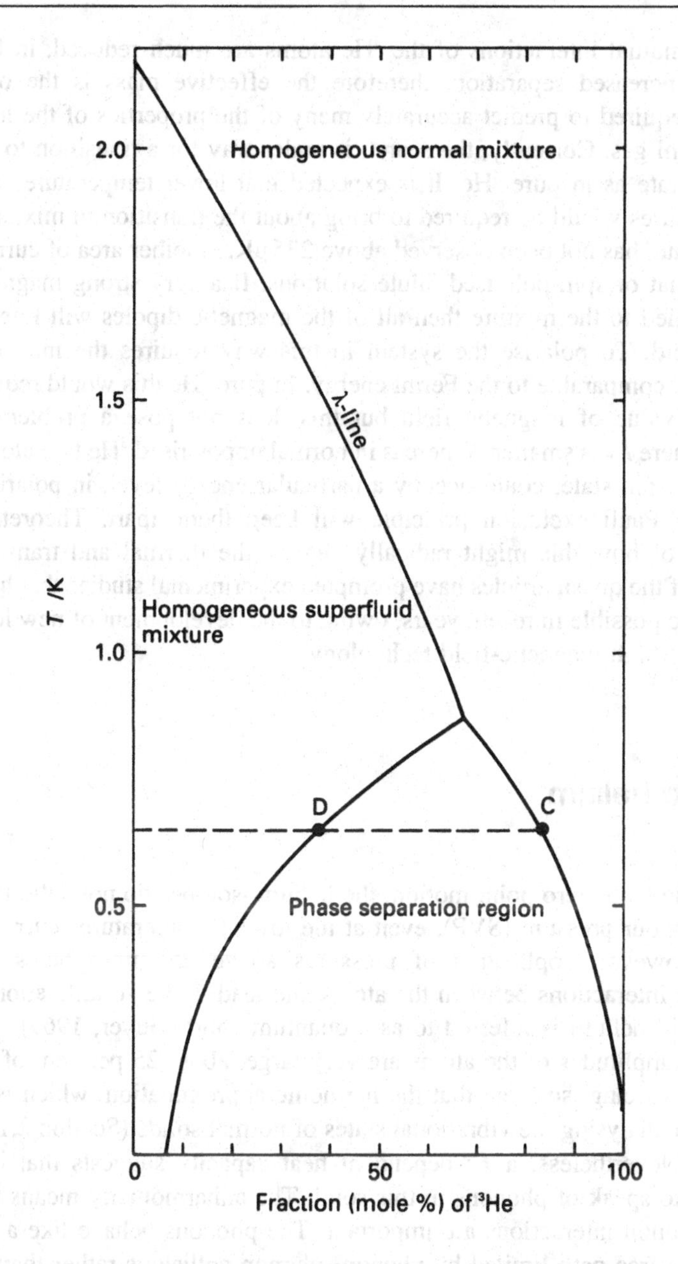

Figure 3.15 *Phase diagram for 3He–^{4}He mixtures [after Taconis, K. W. and De Bruyn Ouboter, R. (1964)* Prog. Low. Temp. Phys., *4, 38]*

temperature. At temperatures less than about 10 mK, the specific heat is linearly dependent on temperature, as is expected for a degenerate Fermi gas. Interactions between the ^{3}He and the ^{4}He atoms give rise to an effective mass of the ^{3}He particles that is some 2.5 times larger than the bare mass of a ^{3}He atom. The mutual interactions of the ^{3}He atoms are much reduced, in line with their increased separation, therefore the effective mass is the only parameter required to predict accurately many of the properties of the low-density Fermi gas. Currently, the search is under way for a transition to the superfluid state as in pure ^{3}He. It is expected that lower temperatures and higher pressures would be required to bring about the transition in mixtures, which, to date, has not been observed above 235 μK. Another area of current interest is that of spin-polarised dilute solutions. If a very strong magnetic field is applied to the mixture then all of the magnetic dipoles will line up with the field. To polarise the system in this way requires the magnetic energy to be comparable to the Fermi energy. In pure ^{3}He this would require too high a value of magnetic field but this does not pose a problem in mixtures where E_F is smaller. Whereas in normal unpolarised ^{3}He two atoms, one of each spin state, could occupy a particular energy level, in polarised systems the Pauli exclusion principle will keep them apart. Theoretical predictions of how this might radically change the thermal and transport properties of the quasiparticles have prompted experimental studies that have only become possible in recent years, owing to the development of new low-temperature–high-magnetic-field technology.

3.4 Solid helium

Owing to the large zero point motion, the helium isotopes do not solidify at saturated vapour pressure (SVP), even at the lowest temperatures currently reached. However, application of pressures above 25 atmospheres can increase the interactions between the atoms and lead to the solidification of helium. Solid helium is referred to as a quantum solid (Guyer, 1969). The vibrational amplitudes of the atoms are very large, about 25 per cent of the interatomic spacing, so large that the harmonic approximation, which is so useful when analysing the vibrational states of normal solids (Section 2.1.4), is invalid. Nevertheless, a T^3-dependent heat capacity suggests that it is reasonable to speak of phonons in this solid. The anharmonicity means that phonon–phonon interactions are important. The phonons behave like a gas with a mean free path limited by phonon–phonon collisions rather than by scattering by the walls as in normal solids at low temperatures. Localisation of helium atoms in the solid means that the quantum statistics applicable to the liquid phase are not relevant to solid. In the case of solid ^{3}He additional entropy can be carried by the nuclear spin system. A consequence of this is that, in a certain range of temperature, the entropy of the solid is higher than that of the liquid. This is another manifestation of the unusual effects observed in quantum systems (see also Section 5.3).

3.5 Adsorbed helium films

If helium gas is allowed to come into contact with any solid surface at a low temperature it will condense onto the surface. The helium atoms are held on the surface by forces of a van der Waals nature. By limiting the amount of gas reaching the surface, a film of helium only one atom thick (a monolayer) can be formed. Motion of the adsorbate atoms in a direction normal to the surface is restricted by the retaining forces, but motion parallel to the surface is allowed. If the surface is flat and there is no lateral variation in the attractive forces, then the adsorber acts only to confine the adsorbate to a plane. Under these conditions, the adsorbed gas behaves as a quasi-two-dimensional system. This system is of considerable interest both theoretically and experimentally; for example, Bose condensation is not predicted to occur in 2-D films of ^{4}He, however, there have been experimental observations of superfluid behaviour in such films (Bishop and Reppy, 1978).

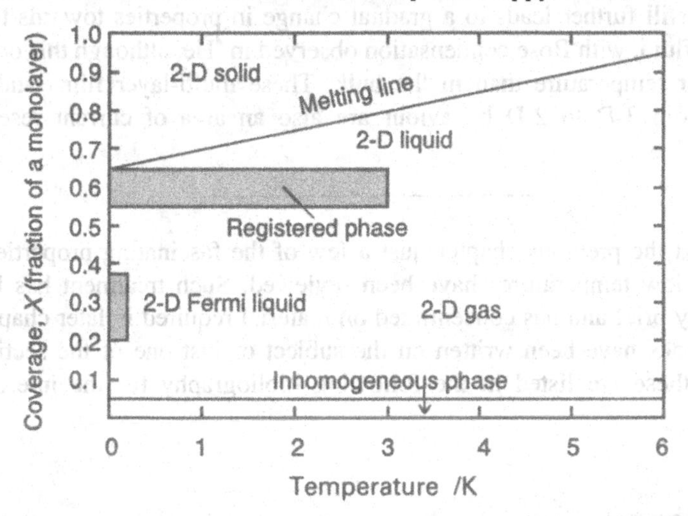

Figure 3.16 *Coverage–temperature phase diagram of sub-monolayer films of ^{3}He adsorbed on exfoliated graphite. Increasing the coverage forces the helium atoms closer together, much as does increasing the pressure in bulk systems. See text for an explanation of the different phases*

A number of possible substrates for use in experimental studies have been investigated (Dash, 1975) and few even approximate to the ideal requirements of large, flat and homogeneous crystalline surfaces. The most successful to date is exfoliated graphite, which consists of many flat graphite crystal platelets, about 10 nm across. Figure 3.16 shows the phase diagram of ^{3}He adsorbed on exfoliated graphite. The coverage, X, is a measure of the film density, $X = 1$ corresponds to the formation of one full monolayer at a density of 2.4×10^{19} helium atoms per m^2. At this density, the average interatomic separation is 0.327 nm as compared to 0.45 nm in the bulk at

SVP, so, a result of the mediation of the substrate, the helium is able to solidify in 2-D without the application of external pressure. At lower coverages, 2-D fluid phases are found with a 2-D Fermi liquid at low temperatures. The registered phase corresponds to helium atoms being trapped by the periodic potential of the substrate on the scale of the substrate's lattice spacing. The helium atoms each occupy one out of three of the carbon hexagons to form a triangular lattice with a spacing of 0.42 nm. At very low coverages, $X < 0.1$, the effects of imperfections in the substrate, crystallite edges, dust etc. become important and dominate the properties of the film. It appears that patches of solid helium nucleate at the imperfections.

Measurements of the adsorbate pressure above the substrate, at constant temperature, as the coverage is increased (adsorption isotherms), show steps which indicate the formation of distinct layers of adatoms up to $X = 4$ or 5. At low densities, the second layer behaves as an imperfect 2-D gas. When some third-layer atoms are present, the second layer solidifies. Increasing the coverage still further leads to a gradual change in properties towards those of a bulk fluid, with Bose condensation observed in ^{4}He, although this occurs at a lower temperature than in the bulk. These multi-layer films and the change from 3-D to 2-D behaviour are also an area of current research interest.

In this and the previous chapter, just a few of the fascinating properties of matter at low temperatures have been reviewed. Such treatment has been necessarily brief and has concentrated on material required in later chapters. Whole books have been written on the subject of just one of the sections; some of these are listed in the following bibliography for the interested reader.

Bibliography

Bishop, D. J. and Reppy, J. D. (1978). *Phys. Rev. Lett.*, **40**, 1727

Dash, J. G. (1975). *Films on Solid Surfaces* (New York: Academic Press)

Guyer, R. A. (1969). 'The physics of quantum crystals', in *Solid State Physics*, Vol. 22, edited by F. Seitz, D. Turnbull and H. Ehrenreich (New York: Academic Press), 413

Landau, L. D. (1957). 'The theory of a Fermi liquid', in *Soviet Physics-JETP*, Vol. 3, 920

McClintock, P. V. E., Meredith, D. J. and Wigmore, J. K. (1984). *Matter at Low Temperatures* (Glasgow: Blackie)

Vinen, W. F. (1961). *Proc. R. Soc. Lond.*, **A240**, 218

Wheatley, J. C. (1975a). 'Three lectures on the experimental properties of liquid ^{3}He', in *The Helium Liquids*, edited by J. G. M. Armitage and I. E. Farquhar (New York: Academic Press) 241

Wheatley, J. C. (1975b). *Rev. Mod. Phys.*, **47**, 466
Wilks, J. (1970). *An Introduction to Liquid Helium* (Oxford: Clarendon)

4

Reaching low temperatures, stage 1: ^{4}He cryogenic systems, 300-1K

The majority of low-temperature experiments in research and in industry are carried out at temperatures down to about 1 K. The techniques used for achieving such temperatures can, with care, be used to reach temperatures as low as 0.3 K. Cooling to temperatures of this order is a necessary first stage to achieving much lower temperatures, using the techniques described in the following chapters.

4.1 Liquid cryogens

The standard techniques for achieving temperatures in the range 300 K to 1 K utilise liquid cryogens. These are obtained by liquefying certain gases. When boiling under atmospheric pressure liquid cryogens provide reliable constant-temperature baths into which the experimental system may be immersed. Cooling to still lower temperatures can be achieved by reducing the vapour pressure above the liquid. Table 4.1 lists the properties of a number of liquid cryogens. Liquid oxygen and hydrogen used to be popular but are rarely used nowadays, owing to the potential hazards. Liquid nitrogen and the common isotope of helium (^{4}He) are the most widely used cryogens. Pumping on ^{4}He enables temperatures down to about 1 K to be reached. The lighter isotope of helium, ^{3}He, is used only to achieve the lowest temperatures in this range; it is very scarce and highly expensive. Liquid-cryogen production and usage will be covered in the remainder of this chapter.

4.2 Liquefaction of gases

There are three methods that may be used to liquefy gases. These are:

1 Direct liquefaction by isothermal compression.

2 Making the gas perform work against external forces at the expense of its internal energy, leading to cooling and eventual liquefaction.

3 Making the gas perform work against its own internal forces by Joule–Kelvin expansion.

The principles underlying each of these techniques now will be considered in turn.

4.2.1 Isothermal compression

A set of idealised P–V isotherms, known as Andrews curves, for a hypothetical gas are shown in Figure 4.1. There are two cases to consider, corresponding to isothermal compression of the gas at temperatures above or below its critical temperature, T_{cp} (see Table 4.1).

Table 4.1 Thermodynamic properties of a number of liquid cryogens

Cryogen	Boiling point @ 10^5 Pa/K	Critical point T_{cp}/K	P_{cp}/kPa	Latent heat of vaporization /kJ l^{-1}	Inversion temperature/K
Hydrogen	20.4	33.2	1 300	30	203
Nitrogen	77.3	126	3 350	160	625
Oxygen	90.2	154.5	5 010	240	762
^{4}He	4.2	5.2	230	2.6	43.2
^{3}He	3.2	3.32	120	0.5	–

If $T > T_{cp}$, compression takes the gas along the curve from A to B, and although the gas density can become comparable to the density of the liquid, no liquid is formed. If, on the other hand, $T < T_{cp}$, compression takes the gas from C to D, where the isotherm crosses the liquid–vapour coexistence curve. Further compression to E causes the gas to become completely converted to liquid. Ammonia ($T_{cp} = 406$ K) is one gas that can be liquefied in this way starting out at room temperature.

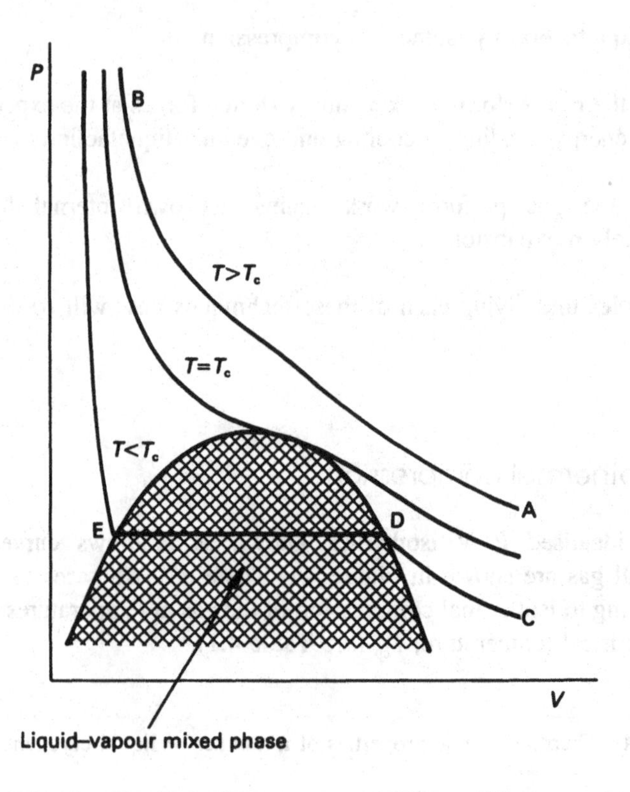

Figure 4.1 *P–V Isotherms (Andrews curves) for a hypothetical gas*

4.2.2 Performance of external work by the gas

If a thermally isolated gas is made to perform external work, it does so at the expense of its internal energy, and its temperature must fall as its thermal energy is expended, whatever the starting conditions of temperature, pressure etc. A well-known example of this process is the steam engine, where superheated steam enters the cylinder, performs work moving the piston and is ejected at a much lower temperature. Sometimes water droplets are seen to form in the exhaust gas as it begins to liquefy.

In practice, the gas is isothermally compressed, which involves removing an amount of heat corresponding to the work done on the gas. The gas is then allowed to expand adiabatically, resulting in cooling. The thermodynamic principles behind this process can readily be demonstrated in the special case of an ideal gas: The first law of thermodynamics gives (see Section 1.2.2)

$$dQ = dE + p\,dV,$$

where dQ is the heat exchanged with the surroundings, $p\,dV$ the work done by the gas, when its volume is changed by amount dV, at constant pressure, and dE the change in internal energy; for one mole of gas

$$dE = c_V\,dT,$$

where c_V is the molar specific heat capacity of the gas at constant volume. In a thermally isolated process,

$$dQ = c_V\,dT + p\,dV = 0.$$

The equation of state of one mole of an ideal gas is $pV = RT$, therefore

$$p\,dV + V\,dp = R\,dT.$$

Eliminating dT between these two equations and rearranging gives

$$\gamma\left(\frac{dV}{V}\right) + \frac{dp}{p} = 0,$$

where $\gamma = (c_V + R)/c_V$. Integrating,

$$\ln V + \ln p = \text{constant},$$

therefore

$$pV^{\gamma} = \text{constant}.$$

This equation relates the pressure to the volume of the gas during the expansion. Since $\gamma > 1$, the gas leaves the expansion stage at low pressure. To relate the temperature to the volume, it is only necessary to use the equation of state for an ideal gas to substitute for p, which gives

$$V^{(\gamma-1)}T = \text{constant}.$$

Now $\gamma - 1 = R/c_V > 0$, hence expansion leads to cooling. For a monatomic ideal gas $R/c_V = 2/3$, and so doubling the volume occupied by the gas leads to a reduction in its temperature by about 40 per cent.

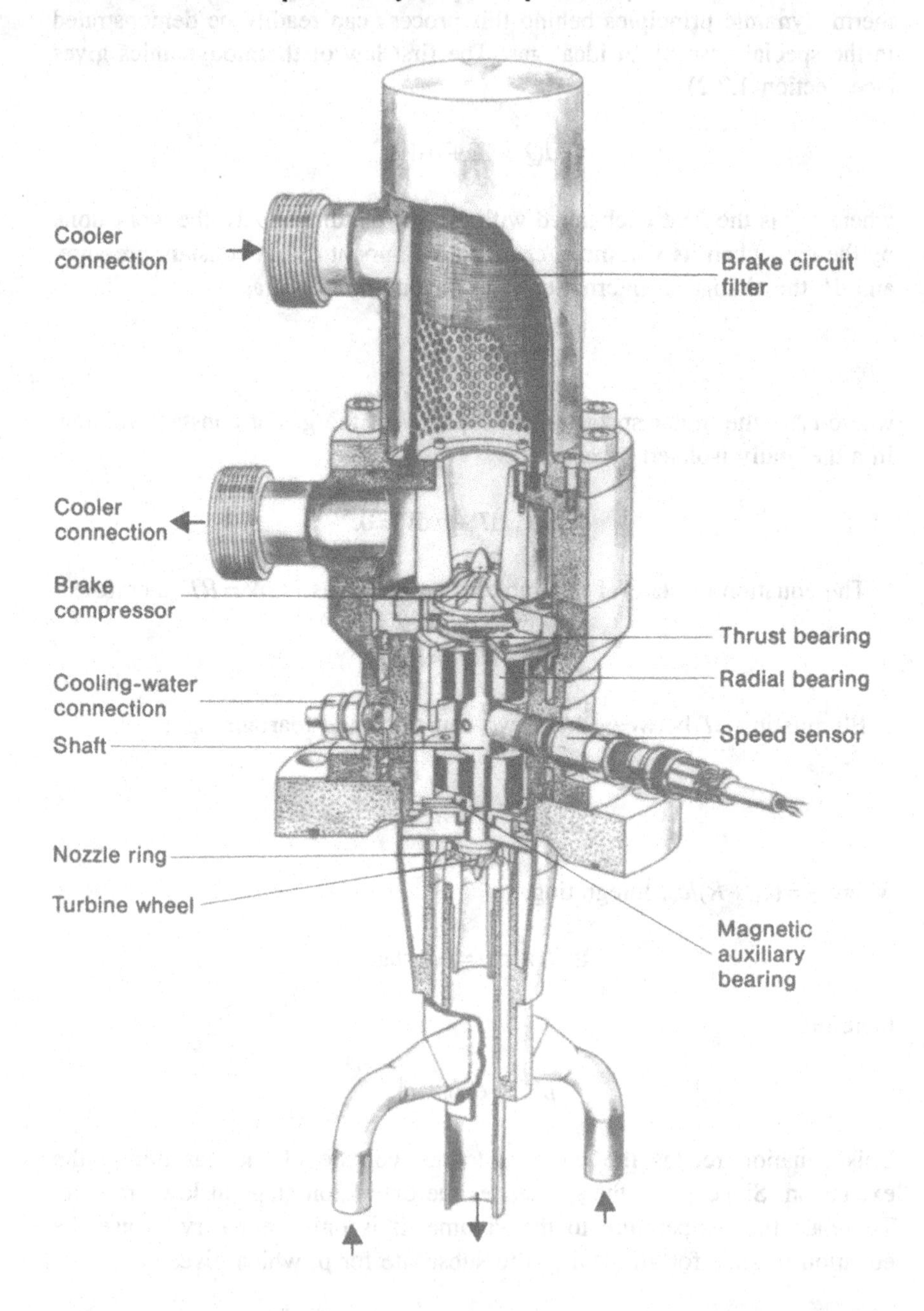

Figure 4.2 *Turbine in which gas is cooled by doing work turning the blades. The work is dissipated in a brake circuit. Gas bearings are used because ordinary lubricants would freeze and seize the mechanism. (Reproduced by courtesy of Linde Cryogenics Ltd.)*

In practice, the gas can be made to operate a valved piston-in-cylinder arrangement, the work being dissipated by dampers connected to the pistons. Alternatively, it may be expanded via a set of turbine blades. These are connected via a shaft to another set of blades which operate as a brake, dissipating the energy in heating another gas (Figure 4.2). At high temperatures this cooling process is quite efficient, but as the liquefaction point is reached c_V and, hence, the efficiency reduce. Another drawback of this method is that it is necessary to have moving mechanical components operating at low temperatures which can pose lubrication problems. Conventional lubricants freeze solid at cryogenic temperatures. Nevertheless, the technique is widely used as the first-stage cooling process in practical helium liquefiers (see Section 4.2.4).

A practical example of a piston-in-cylinder expansion engine that may be used as a liquefier is the Stirling refrigerator. This is basically a Stirling heat engine operating in reverse. The gas is actually expanded isothermally rather than adiabatically. Figure 4.3 shows a schematic representation of the Stirling refrigerator and the operating cycle. The engine consists of two cylinders, one at high temperature, T_h, and one at low temperature, T_l. These are connected via a regenerator which should ideally possess the following qualities:

1 negligible resistance to gas flow;

2 large, internal, surface area to ensure maximum heat transfer between the gas and the regenerator;

3 large, specific, heat capacity at low temperatures;

4 small volume so as not to store large amounts of gas; and

5 very small or zero, longitudinal, thermal conductivity to minimise direct heat leak between the hot and cold cylinders.

All of these, with the exception of 3 (see Section 2.1.3), are fairly easy to achieve in practice, though some compromise is necessary to satisfy 2, 3 and 4 simultaneously.

The four-step cycle proceeds as follows:

1 The piston in the hot cylinder moves to compress the gas isothermally. Heat Q_h is released at this stage. The lower piston is held fixed.

2 The lower piston is released and the upper piston is moved to transfer the gas to the lower, low-temperature cylinder. This step takes place at constant volume. Assuming that the system has been operating for some time, the regenerator will already be at a lower temperature than the gas. Consequently, as the gas passes through the regenerator, it is cooled and enters the lower cylinder at reduced pressure and temperature.

3 The lower piston is moved to expand the gas which, to maintain its constant temperature, extracts heat Q_l from the low-temperature part of the system.

4 Finally, the lower piston is moved to return the gas at constant volume to the hot cylinder. The gas passes back through the regenerator cooling it on the way. The cycle is then repeated.

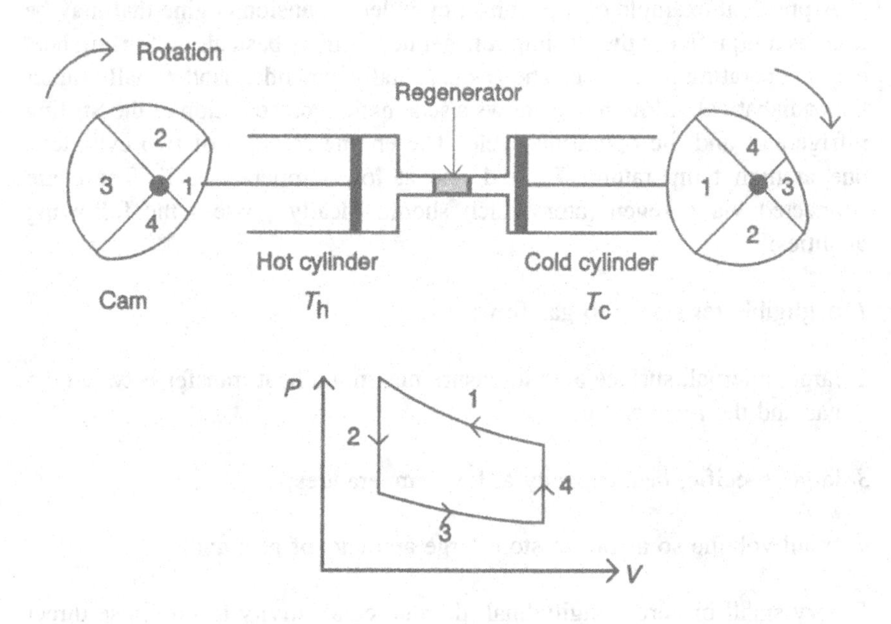

Figure 4.3 *Schematic diagram of Stirling refrigerator and P—V diagram for the four-stage refrigeration cycle*

Because it is difficult to make regenerators with a sufficiently high specific heat capacity near to 4.2 K, it is hard to liquefy helium using the Stirling refrigerator. However, by using helium gas as a refrigerant, it is possible to make small, closed cycle refrigerators capable of achieving base temperatures near to 10 K. To produce liquid nitrogen, large condensing surfaces are attached to the cold cylinders over which a flow of nitrogen gas is passed. Alternatively, the small quantities of liquid that collect in the regenerator may be drained off and fresh gas admitted to the hot cylinder.

4.2.3 Isenthalpic or Joule–Kelvin expansion

The basic idea behind this method is to make the gas do work against its own internal forces. The advantages are simplicity and the absence of low-temperature moving parts. In practice, the gas is expanded via a small nozzle or porous plug thermally isolated from the surroundings. For a real gas, isenthalpic expansion results in cooling, if the gas is initially below its inversion temperature.

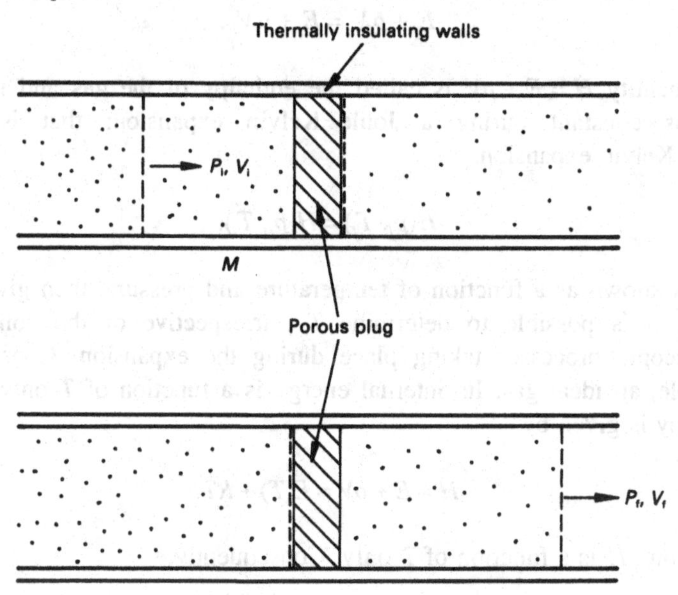

Figure 4.4 *Joule–Kelvin expansion of mass* M *of a gas through a porous plug*

Consider the expansion of mass M of gas, initially at pressure p_i and volume V_i, through a porous plug, as shown in Figure 4.4. If we assume that the volume of the plug is small compared with V_i then nearly all of the gas emerges on the other side of the plug at pressure p_f and volume V_f. The internal energy of a gas can be expressed as a function of its pressure and temperature. The change in internal energy as a result of the expansion is given by

$$\Delta E = E_f - E_i = E(p_f, T_f) - E(p_i, T_i).$$

The process also involves work. Work p_iV_i must be done on the gas on the input side and the gas does work p_fV_f on the output side. Hence

$$\Delta W = p_fV_f - p_iV_i.$$

In an adiabatic process, where the heat Q exchanged with the surroundings is zero, application of the first law of thermodynamics to the mass of gas gives

$$(E_f - E_i) + (p_f V_f - p_i V_i) = 0.$$

Rearranging,

$$E_f + p_f V_f = E_i + p_i V_i .$$

The quantity $H = E + pV$ is called the enthalpy of the gas and it clearly remains constant during a Joule–Kelvin expansion; that is, for a Joule–Kelvin expansion,

$$H(p_f, T_f) = H(p_i, T_i).$$

If H is known as a function of temperature and pressure, then given p_i, T_i and p_f it is possible to determine T_f, irrespective of the complicated microscopic processes taking place during the expansion. Consider, for example, an ideal gas. Its internal energy is a function of T only and the enthalpy is given by

$$H = E + pV = E(T) + RT.$$

Therefore, H is a function of T only. Consequently,

$$H(T_f) = H(T_i),$$

or $T_f = T_i$, proving that the temperature of an ideal gas does not change during an isenthalpic or Joule–Kelvin expansion. This is to be expected, as there are no internal forces for the gas to do work against.

In the more general case of a real gas, from a knowledge of $H(T, p)$, it is possible to construct curves of T versus p for various fixed values of H. Such curves are known as isenthalps. Figure 4.5 shows a set of isenthalps for nitrogen. Over a certain range of temperature and pressure the curves exhibit maxima. The locus of these maxima is called the inversion curve. For a starting point to the left of the inversion curve, the expansion process always results in cooling. For a starting point to the right of the inversion curve, the expansion can result in heating or cooling, depending on the finishing point. For a starting temperature $T > T_{inv}(\max)$, expansion invariably results in heating. Clearly the optimum starting pressure for a given finishing pressure will lie on the inversion curve. The effectiveness of the process depends on the slope, $\mu = (\partial T / -\partial p)_H$, of the isenthalps, known as the Joule–Kelvin coefficient.

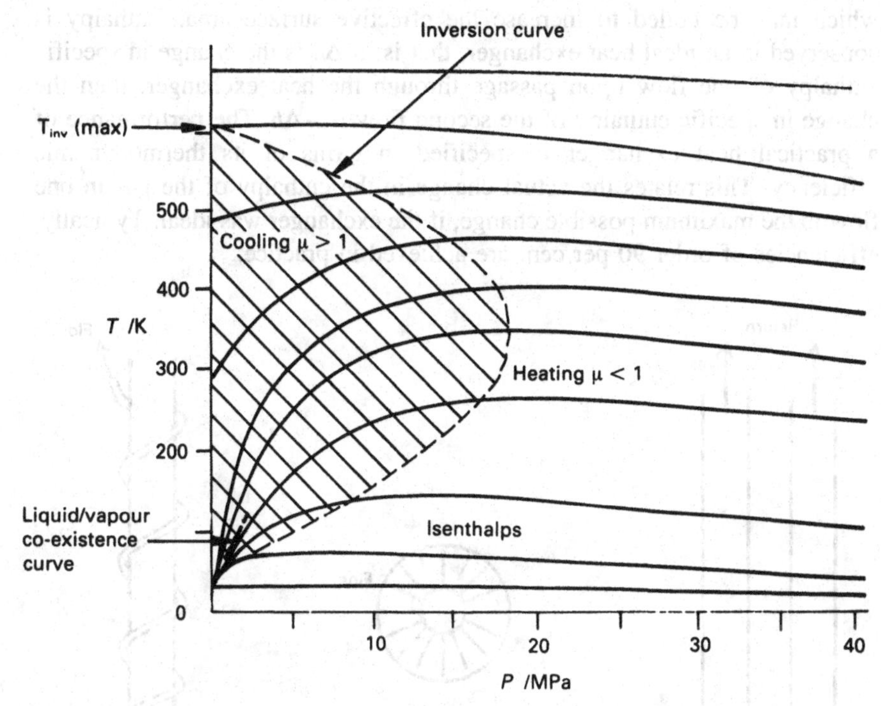

Figure 4.5 *P–T Isenthalps for nitrogen gas (after Reif, 1965)*

At very low temperatures, some of the isenthalps cross the liquid–vapour equilibrium curve at which some of the gas liquefies. Nitrogen with T_{inv}(max) = 607 K can be liquefied in this way, starting out at room temperature. Helium, however, needs to be pre-cooled to below 43 K before the Joule–Kelvin effect can be used (see Section 4.2.4).

The cooling produced by a Joule–Kelvin expansion is generally insufficient to liquefy directly from the high-pressure gas supply. In practice, it is necessary to use the cold low-pressure gas produced in the initial expansion to pre-cool the incoming gas. After a while, the temperature of the gas reaching the expansion nozzle is low enough for liquid to be produced in the expansion. This pre-cooling is achieved with the aid of counterflow heat exchangers (see Figure 4.6), which should ideally possess the following properties:

1 a minimum flow resistance to incoming and outgoing gas;

2 a large surface area for heat transfer between the two flows, and

3 small thermal mass to minimise cool down time.

These properties are achieved fairly easily in practice using concentric tubes, which may be coiled to increase the effective surface area. Enthalpy is conserved in an ideal heat exchanger, that is, if Δh is the change in specific enthalpy of one flow upon passage through the heat exchanger, then the change in specific enthalpy of the second flow is $-\Delta h$. The performance of a practical heat exchanger is specified in terms of its thermodynamic efficiency. This relates the actual change in the enthalpy of the gas in one flow to the maximum possible change, if the exchanger was ideal. Typically, efficiencies of order 90 per cent are achieved in practice.

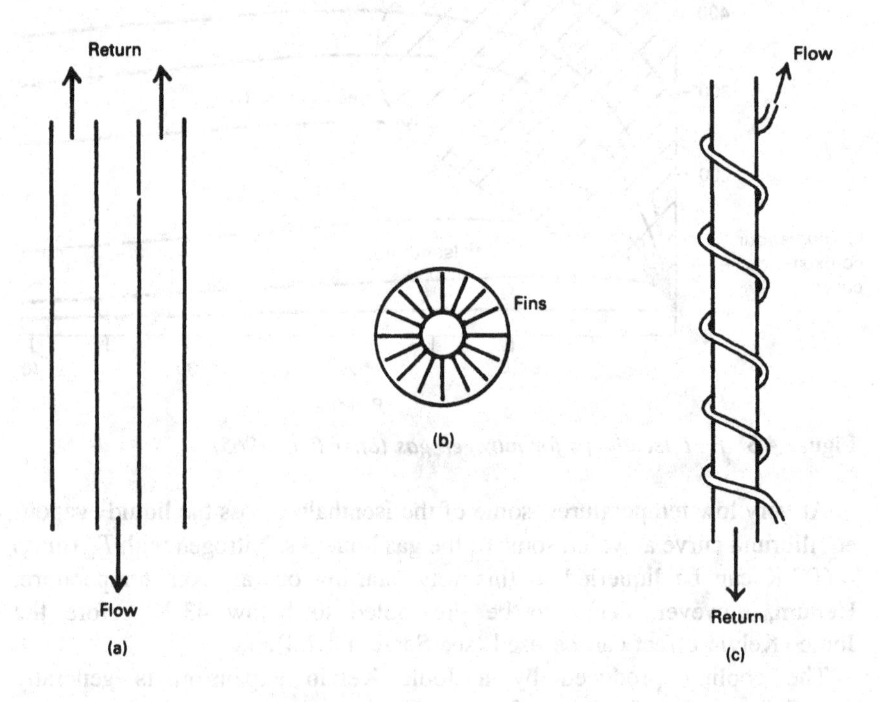

Figure 4.6 *(a) Simple counterflow heat exchanger using concentric tubes. (b) Increasing the surface area by adding fins to the tube. (c) Coiling the tube to increase efficiency*

A Joule–Kelvin expansion nozzle preceded by a counterflow heat exchanger forms the basis of the simple Hampson air or nitrogen liquefier shown in Figure 4.7. The compressed input gas is passed first through a purifier to remove any contaminants, such as water vapour, which might freeze and block the expansion orifice. Liquid is collected in a thermally insulated container. Total enthalpy and mass are conserved in an ideal Hampson liquefier, hence

$$m_i h_i = m_o h_o + m_l h_l ,$$

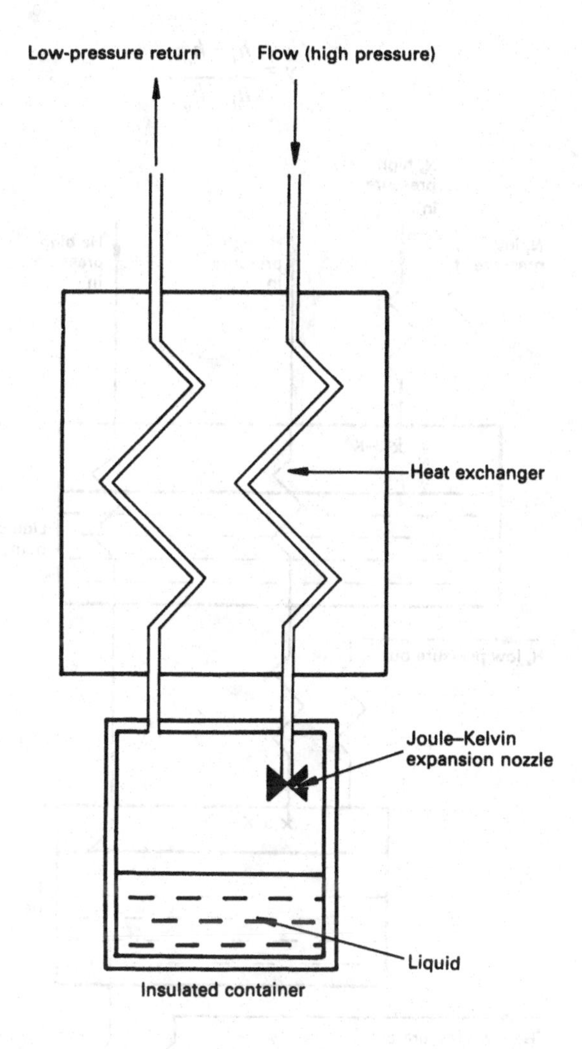

Figure 4.7 *Hampson air or nitrogen liquefier consisting of a Joule–Kelvin expansion stage and a counterflow heat exchanger*

and

$$m_i = m_o + m_l\,,$$

where h_i, h_o and h_l are respectively the specific enthalpies of the inlet gas, outlet gas and liquid; m_i and m_o are respectively the mass flow rates of the inlet and outlet gas and m_l is the mass of liquid collected per unit time. It is, therefore, straightforward to show that the yield of the liquefier, defined as $y = m_l/m_i$, is given by

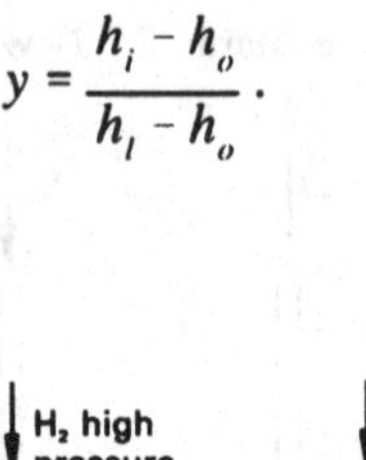

$$y = \frac{h_i - h_o}{h_l - h_o} .$$

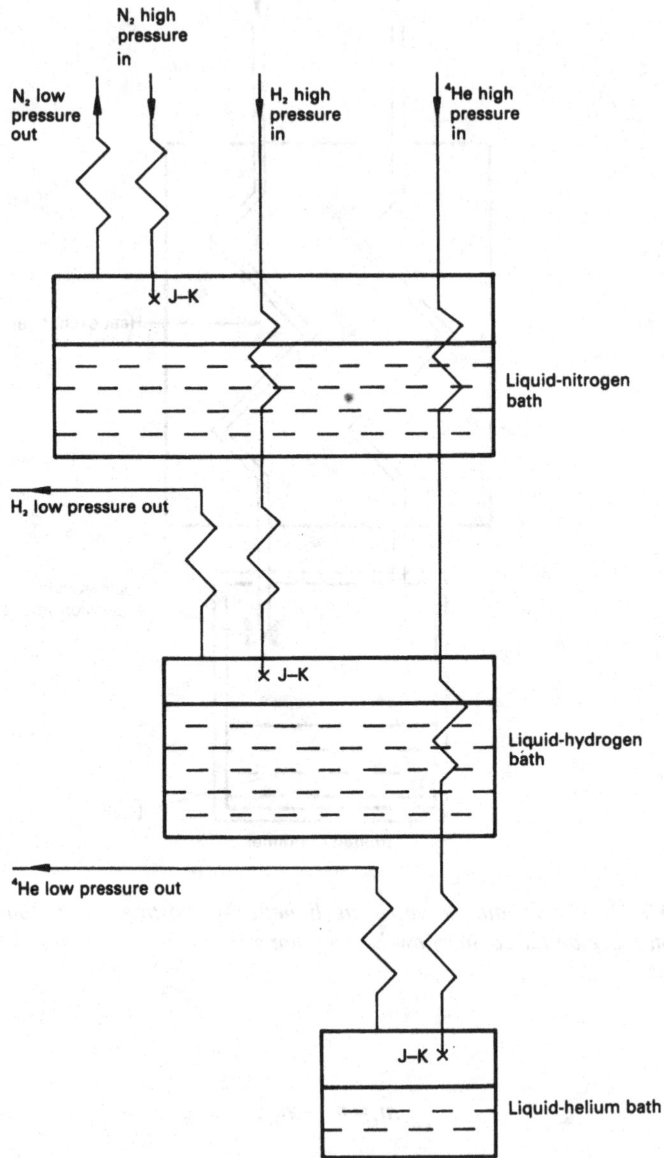

Figure 4.8 *Schematic diagram of cascade process helium liquefier using liquid-nitrogen and liquid-hydrogen baths for pre-cooling*

4.2.4 A practical helium liquefier

A combination of the techniques described in Sections 4.2.2 and 4.2.3 is used in order to liquefy helium. Helium can be liquefied by a Joule–Kelvin expansion, but it is necessary to pre-cool the gas to (or to below) its maximum inversion temperature, 43 K, prior to the expansion. One method of pre-cooling involves passing the gas through, and allowing it to exchange heat with, baths of more easily liquefied cryogens, the so-called cascade process. An example of this is shown in Figure 4.8. The first stage is to pass the helium gas through a liquid nitrogen bath at 77 K (nitrogen can be liquefied by Joule–Kelvin expansion starting from room temperature). The helium at near to 77 K is then passed through a liquid hydrogen bath (the hydrogen is liquefied by Joule–Kelvin expansion starting at 77 K). The emerging helium, at a temperature of about 20.4 K, is then liquefied using a Hampson-type system.

Helium liquefiers based on the cascade process require three separate gas supplies and are rather cumbersome. Furthermore, there are obvious hazards associated with the use of hydrogen. In modern helium liquefiers, the helium gas is pre-cooled by making it do mechanical work, either in a piston-in-cylinder arrangement or via a turbine. The first liquefier using mechanical pre-cooling to be put into commercial production was the A. D. Little–Collins liquefier. It was to revolutionise the development of low-temperature physics from the mid-1940s because of its compactness and convenience of operation. Although there have been improvements in technology leading to better thermodynamic efficiency, reliability and production rates, the basic operating principles of helium liquefiers have changed little.

Figure 4.9 shows a schematic diagram of the A. D. Little–Collins liquefier system, which operates as follows: The impure helium gas returned from the experimental systems is stored in a low-pressure gasometer. The gas is then compressed via a purifier into a medium pressure store. To replace any losses, new gas may be added from supply cylinders prior to the purification stage. The purifier removes contaminants such as air, which will freeze and might cause the mechanical parts of the liquefier to seize. In its simplest form, it consists of liquid-nitrogen-cooled tubes containing activated charcoal. As the helium passes through the tubes, any traces of impurities are adsorbed by the cold charcoal. The liquefaction process starts by compressing the pure helium to about 1.4×10^6 Pa (14 atmospheres) in a multi-stage compressor. The gas then enters the first counterflow heat exchanger, where it is cooled to around 60 K by the low-pressure gas exiting from the liquefier. The gas flow is then split; approximately 70 per cent enters the next heat exchanger and the remaining 30 per cent enters the first expansion engine, in which it is cooled to around 30 K. After passage through a third heat exchanger, the gas flow is split again and about 55 per cent of the gas enters the second expansion engine working between 15 K and 8 K. The remaining helium then passes through the final heat exchanger to the Joule–Kelvin expansion stage. The expansion nozzle is sometimes located at the end of an insulated heat exchanger tube which can be inserted directly into mobile helium storage Dewars. This avoids wasteful transfers of liquid between containers. The heat exchanger tube consists of three concentric pipes, the inner two forming a counterflow heat exchanger and the outer one holding a vacuum to thermally isolate the heat exchanger from the surroundings. The exhaust gas from the engines and the expansion stage leaves the system at low pressure and is returned to the multi-stage compressor to be recycled through the system.

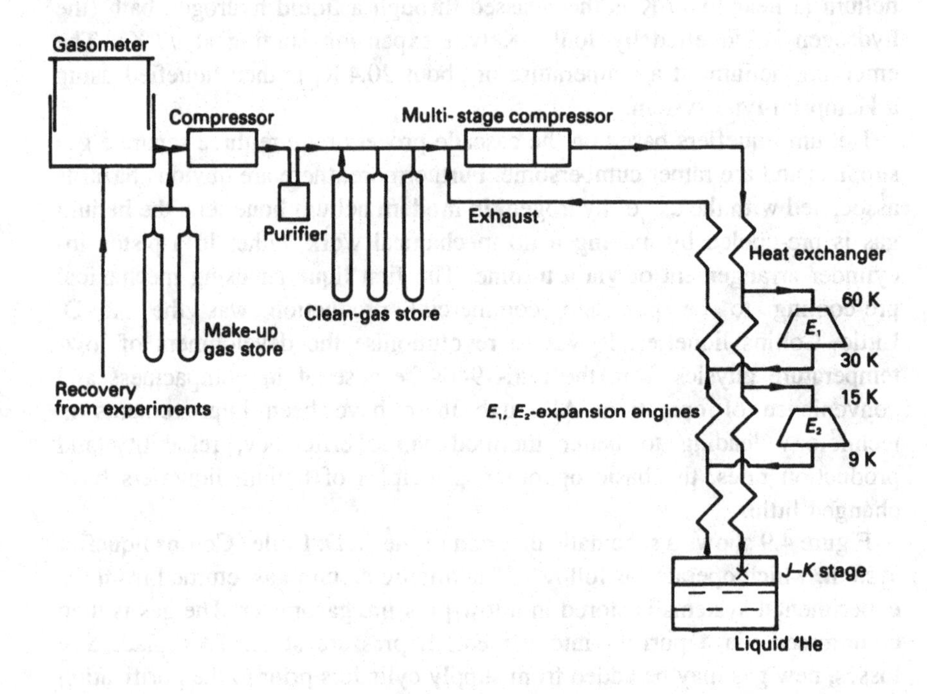

Figure 4.9 *A. D. Little–Collins helium liquefier and gas handling systems*

Enthalpy is conserved in the heat exchangers and the Joule–Kelvin expansion stage, so the mass yield of the Collins system can be written in terms of the specific enthalpies of the high-pressure input gas, h_i, the low-pressure output gas, h_o, and the liquid, h_l,

$$y = \frac{h_i - h_o + x_1 \Delta h_1 + x_2 \Delta h_2}{h_l - h_o},$$

where x_1 and x_2 are the fractional mass flows through each expansion engine stage in which the enthalpy changes are Δh_1 and Δh_2 respectively. Mass is conserved in the system as a whole, therefore,

$$y + m_0 + m_i = 1.$$

The typical mass yield of the two-stage process described above is about 3.5 per cent. Therefore, in order to produce 1 litre of liquid, 23 000 litres (STP) of gas must be circulated through the system. Various methods are used to improve the yield. Pre-cooling with liquid nitrogen can increase it to around 10 per cent, or extra expansion engine stages can be used.

4.3 Storage and handling of liquid cryogens

Helium liquefiers are large pieces of apparatus and are often removed from the experimental systems in which the helium is used. It is, therefore, essential to be able to store and transport the liquid. Helium storage vessels are designed to minimise liquid boil-off due to heat leaking into the storage space. Vacuum-insulated metal containers, i.e. Dewars, are used (Figure 4.10). The liquid and gas is extracted via a thin neck which is made of a low-thermal-conductivity alloy, e.g. stainless steel, to reduce heat leaks by conduction. The vacuum jacket surrounding the helium container is very efficient, because any trace of gas is adsorbed by the cold wall of the helium storage space. However, the heat leak due to radiation between two surfaces, one at room temperature and the other at 4.2 K can be very large. To limit this, older storage vessels had a liquid-nitrogen shield around the helium Dewar. Modern vessels use superinsulation instead of the nitrogen jacket. Superinsulation consists of many alternate layers of aluminised Mylar film and low-density fibrous material in a cavity which is pumped to a pressure of around 10^{-1} Pa. Containers using superinsulation are lighter and more convenient because they do not need topping up with nitrogen, but they are not quite as efficient. Typical loss rates for a modern 50-litre dewar are 1–2 litres per day as compared with 1 litre per day for a Dewar with a liquid-nitrogen jacket.

It is not possible to see the liquid level in metal Dewars, so certain methods are required to measure the quantity of liquid stored. Modern depth gauges make use of thin superconducting wires which dip into the helium. (Figure 4.11). The wire is chosen to have a superconducting transition

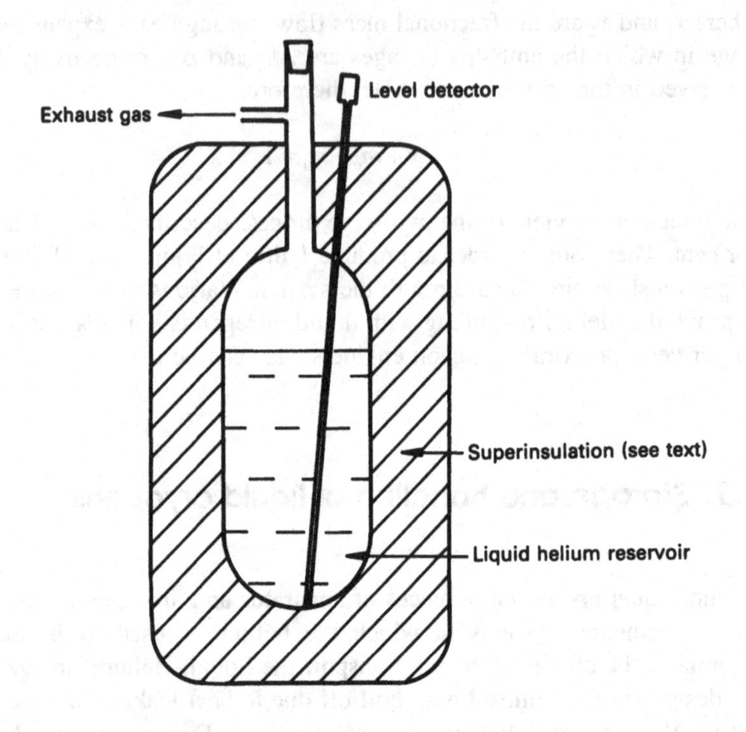

Figure 4.10 *Schematic diagram of a modern stainless steel superinsulated helium Dewar*

temperature a little above 4.2 K. The length of wire below the surface of the liquid is very effectively cooled to below its transition temperature and is superconducting. The length in gas is less effectively cooled and, if a sufficiently large current is passed, it is driven into the normal state. Fortunately, many superconducting materials have a high resistance when normal, and so it is fairly easy from measurements of resistance to determine what fraction of the length of the wire is immersed. Continuous measurement by this technique is inadvisable, because it will cause excessive boil-off of the helium. One advantage of the method is that the superconducting sensor can be left in the Dewar at all times and the current switched on only when a measurement is required. A simple check of the liquid remaining in a storage Dewar may be made by using a gas meter to monitor the volume of gas that has boiled off since the last measurement. 1 litre of liquid helium corresponds to about 0.77 m^3 (or 26.5 cubic feet, since most gas meters are calibrated in imperial units) of gas at room temperature.

Helium gas is an expensive commodity at the time of writing. The price corresponds to £5.00 per liquefied litre, and so it is sensible to recover the gas that boils off from storage Dewars and experiments. Recovery systems require a network of pipes running back to the low-pressure gas store that feeds the liquefier. Some laboratories, whose helium consumption is small, do not liquefy their own gas. In these cases, the helium may be collected in a rubberised bag and then compressed into cylinders for return to the liquid supplier. Care must be taken to ensure the purity of the recovered gas. Various types of one-way valve are used to close the return line and prevent air entering when no helium gas is flowing.

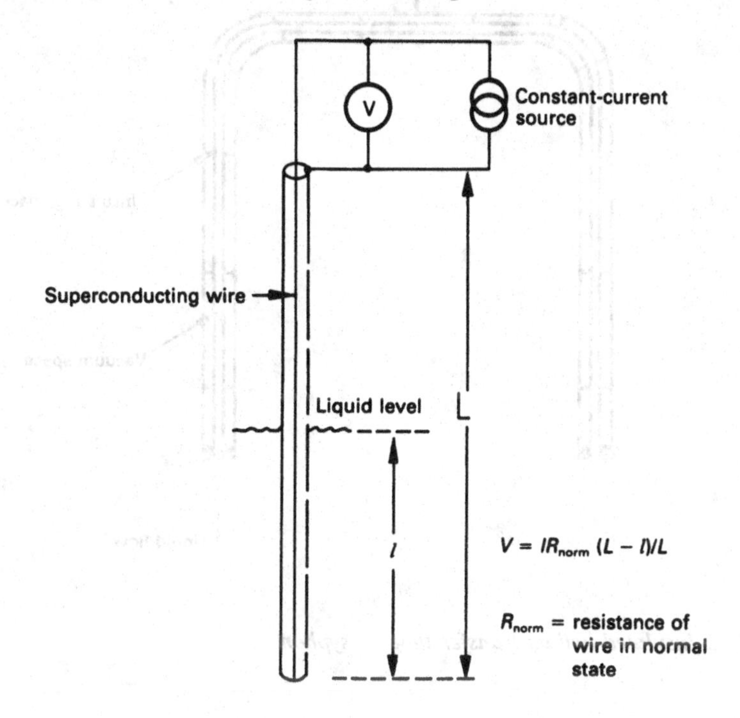

Figure 4.11 *Superconducting liquid-helium depth sensor*

Once the helium has been transported to the experimental system, it has to be transferred from the storage Dewar into the low-temperature apparatus. This is achieved with the aid of an insulated tube, known as a syphon. It consists of two coaxial stainless steel tubes with a vacuum between them (Figure 4.12). One end of the assembly is inserted in the Dewar and the other end into the experimental apparatus. Liquid helium is forced through the inner tube by a slight pressure difference in favour of the Dewar. One method of producing the pressure difference is to attach a small rubber bag, e.g. a football bladder, to the gas space above the liquid helium in the Dewar. By squeezing the bag, warmer gas is forced towards the liquid, thus increasing the boil-off and hence the pressure in the Dewar.

The storage of liquid nitrogen is a much simpler task; a single-stage Dewar or superinsulated vessel can hold the liquid with very little loss for many days. Short-term storage in expanded-polystyrene containers is also possible. Nitrogen is most simply transferred between containers by pouring, or it may be transferred under pressure.

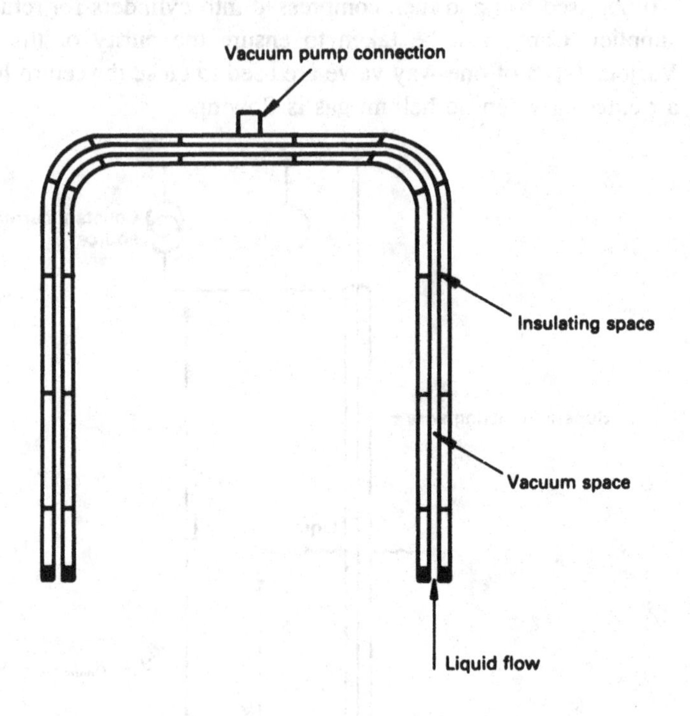

Figure 4.12 *Insulated helium transfer tube or syphon*

4.4 Low-temperature experimental apparatus

A piece of apparatus in which low-temperature experiments are carried out is called a *cryostat*. There are many types of cryostat, each designed specifically for the type of experiment to be performed. Some have windows allowing optical access, or especially thin tails for insertion into the poles of an electromagnet. The experimental specimen may be cooled by direct immersion in the bath of liquid cryogen. Alternatively, it may be mounted in a vacuum and cooled by conduction from the constant-temperature bath. Some cryostats use a continuous flow of cold helium gas to cool the sample. However, in all cases, in designing a cryostat it is important to minimise unwanted heat leaks, which result in the wastage of valuable liquid and restrict the minimum achievable base temperature. The outer casing of a typical helium cryostat (Figure 4.13) usually takes the form of a single superinsulated Dewar or a double Dewar with a liquid-nitrogen jacket. Some

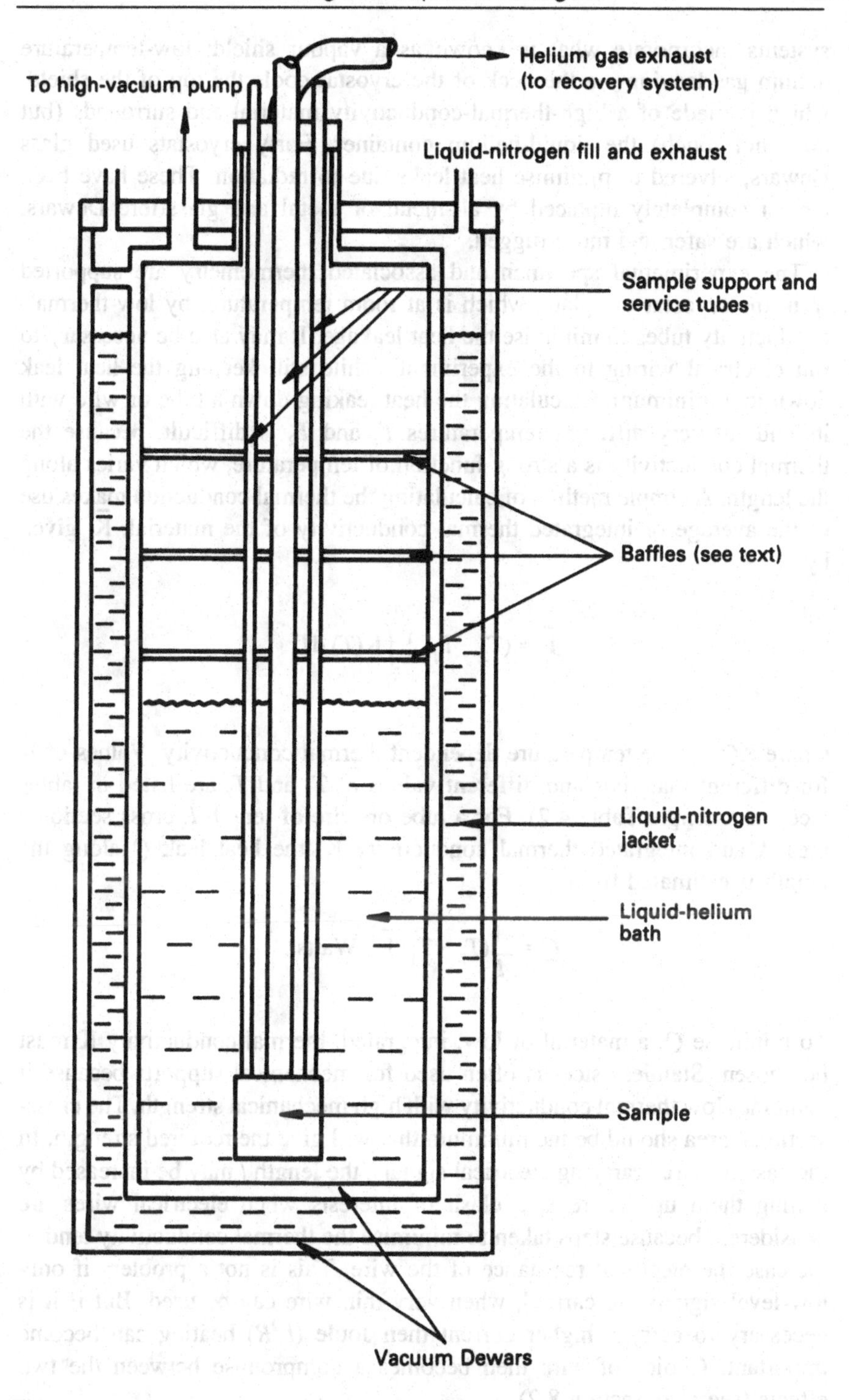

Figure 4.13 *Simple helium immersion cryostat*

systems incorporate what is known as a vapour shield: low-temperature helium gas leaving via the neck of the cryostat cools the top of the shield, which is made of a high-thermal-conductivity material and surrounds (but does not touch) the liquid-helium container. Early cryostats used glass Dewars, silvered to minimise heat leaks due to radiation. These have been almost completely replaced by all-metal or metal and glassfibre Dewars, which are safer and more rugged.

The experimental specimen and associated thermometry are supported from the cryostat top plate, which is at room temperature, by low-thermal-conductivity tubes to minimise the heat leakage. It may also be necessary to run electrical wiring to the experiment, while still keeping the heat leak down to a minimum. Calculating the heat leaking down a tube or wire with its ends at very different temperatures T_1 and T_2 is difficult, because the thermal conductivity is a strong function of temperature, which varies along the length. A simple method of calculating the thermal conduction makes use of the average or integrated thermal conductivity of the material, $\bar{K}$, given by

$$\bar{K} = (T_2 - T_1)^{-1} \int_{T_1}^{T_2} K(T)\, dT,$$

where $K(T)$ is the temperature dependent thermal conductivity. Values of $\bar{K}$ for different materials and different values of T_1 and T_2 are listed in tables (see for example Table 4.2). For a tube or wire of length l, cross sectional area A and integrated thermal conductivity $\bar{K}$, the heat leak $\dot{Q}$ along the length is estimated from

$$\dot{Q} = \frac{A}{l}(T_2 - T_1)\,\bar{K} \text{ Watts}.$$

To minimise $\dot{Q}$, a material of low, integrated, thermal conductivity $\bar{K}$ must be chosen. Stainless steel is often used for mechanical supports because it combines low thermal conductivity with high mechanical strength. The cross-sectional area should be the minimum that will give the required strength. In the case of wires carrying electrical signals, the length l may be increased by coiling them up. There is a clash of interests when electrical wires are considered, because steps taken to minimise the thermal conductivity tend to increase the electrical resistance of the wire. This is not a problem if only low-level signals are carried, when very thin wire can be used. But if it is necessary to carry a higher current then Joule (I^2R) heating can become important. Choice of wire then becomes a compromise between the two effects (see also Section 8.2).

Table 4.2 Integrated thermal conductivities for materials in common cryogenic use

Material	$\bar{K}$/W m^{-1} K^{-1} 300 K–77 K	$\bar{K}$/W m^{-1} K^{-1} 77 K–4.2 K	$\bar{K}$/W m^{-1} K^{-1} 4.2 K–1 K	$\bar{K}$/W m^{-1} K^{-1} 300 K–4.2 K
Copper	410	980	200	551
Brass	81	26	1.7	67
Stainless steel	12.3	4.5	0.2	10
Nylon	0.31	0.17	0.006	0.27

■ EXAMPLE 1

An experiment immersed in liquid helium at 4.2 K requires an electrical current that is supplied via a pair of copper wires each 1 m in length, 0.1 mm diameter and having a total resistance of 0.3 Ω. At what current is the heat input to the helium, due to Joule heating of the wires, equal to the heat input, due to thermal conduction, from the top of the wires at 300 K? What is the helium boil-off rate, in litres per hour, at this current? What would be the effect on the boil-off rate of (a) doubling the diameter of the wires and (b) halving the diameter of the wires?

The heat input due to conduction is given by

$$\dot{Q} = 2A\,(T_2 - T_1)\bar{K},$$

where the cross sectional area of the wire, A is 7.85×10^{-9} m^{-2}. Using the integrated thermal conductivity data in Table 4.2,

$$\dot{Q} = 2.56 \times 10^{-3}\ \text{W}.$$

Heat input due to Joule heating is just $I^2R = 0.3I^2$ watts, and equal to the heat conduction when

$$I = (\dot{Q}/0.3)^{1/2} = 92\ \text{mA}.$$

Total heat input to the helium is then 5.12 mW. The latent heat of vaporisation of helium is 2600 J l^{-1}, and so the boil-off rate is 2 μl s^{-1} or, 7

$cm^3 h^{-1}$.

(a) If the diameter of the wires is doubled then the cross-sectional area is increased by a factor of four. The resistance of the wires is inversely proportional to the cross-sectional area and so the Joule heating is also reduced by the same factor. However, the heat leak due to thermal conduction is quadrupled. The total boil-off rate is increased to $15\,cm^3\,h^{-1}$.

(b) Halving the diameter reduces the cross-sectional area and also the thermal conduction, by a factor of four, but quadruples the Joule heating. Therefore, the total boil-off is again increased to $15\,cm^3\,h^{-1}$.

The heat leak down the tubes and connecting wires can be reduced further by making use of the cold gas evaporated from the liquid to cool them as it leaves the cryostat. Metal plates, called baffles, are attached to the tubes and exchange heat with the cold gas (Figure 4.13). The cooling power of the gas is very high. The heat absorbed by one mole of helium gas being warmed from T_1 to T_2, at a constant pressure of one atmosphere, is given by

$$Q = c_p(T_2 - T_1),$$

where it is assumed that helium behaves as an ideal gas over most of the temperature range. The specific heat capacity of an ideal gas at constant pressure $c_p = 5R/2$, therefore Q is given by

$$Q = \frac{5R\,(T_2 - T_1)}{2}.$$

For $T_2 = 300\,K$ and $T_1 = 4.2\,K$, $Q = 6.15\,kJ\,mol^{-1}$, or $200\,J\,cm^{-3}$. This is high compared with the latent heat of vaporisation of only $80\,J\,mol^{-1}$ or $2.6\,J\,cm^{-3}$.

Thermal conduction is not the only source of unwanted heat input to the low-temperature parts of the cryostat. Radiated heat can also cause considerable problems. According to Stefan's law, the total power radiated per unit area of a body at temperature T is

$$P_r = \varepsilon \sigma T^4,$$

where ε is the emissivity of the surface and $\sigma = 5.7 \times 10^{-8}\,W\,m^{-2}\,K^{-4}$ is the Stefan–Boltzmann constant. For a black body $\varepsilon = 1$, but a highly polished metal surface can have ε as low as 0.01. If the radiation falls on another body, some is absorbed and some is reflected. The rate of energy transfer between two plane parallel surfaces of emissivity ε_1 and ε_2 in equilibrium, at temperatures T_1 and T_2 respectively, is given by

$$P_r = \sigma(T_2^4 - T_1^4)\left(\frac{\varepsilon_1\varepsilon_2}{\varepsilon_1 + \varepsilon_2 - \varepsilon_1\varepsilon_2}\right) \mathrm{W\,m^{-2}}.$$

Polishing the internal surfaces to reduce ε can significantly reduce P_r. For example, the heat transfer between surfaces at room temperature and 4.2 K is 24.3 W m^{-2} if their emissivity is 0.1, and 2.32 W m^{-2} if their emissivity is 0.01 (this is why the interior of a cryostat is not painstakingly polished just to look nice!). If a shield at $T = 77$ K, is inserted between the two surfaces, these figures are considerably reduced to 0.1 W m^{-2} and 10 mW m^{-2} respectively.

Another source of heat leaks, which can be very serious if care is not taken, is the oscillations that can start up within the helium gas, in open-ended tubes running from low to high temperature. These oscillations can be used to measure the depth of liquid helium in opaque containers (Rose-Innes, 1973) and may carry considerable amounts of heat to the low-temperature regions of the cryostat. One way to overcome this problem is to drill small holes along the side of any tubes which could conceivably cause a problem. If this is not possible then putting baffles or some other partial obstruction into the tube may prevent the oscillations.

In addition to minimising heat leaks, it is also important in designing the system to aim for the minimum mass to reduce the quantity of helium required for the initial cool-down. Pre-cooling the system with liquid nitrogen beforehand leads to considerable savings of the more expensive cryogen. However, it is very important to ensure that all traces of liquid nitrogen are removed prior to transferring helium. This is because liquid nitrogen has a relatively large heat capacity and considerable quantities of helium will be wasted in trying to cool it.

Another property of materials that can have a considerable influence on cryostat design is thermal contraction. The effects can be surprisingly large; a cryostat 1 m long can shorten by as much as 2 mm on cooling down. Special care must be taken when a number of different materials are used in the construction of a cryogenic component; the differential contraction can lead to distortion or even failure of the structure.

4.5 Obtaining temperatures above 4.2 K

It is often required to perform experiments at temperatures higher than the normal boiling point of ^{4}He, say between 4.2 and 77 K. Various techniques based on the liquid-helium cryostat have been developed for this. The basic idea is to put the sample in weak thermal contact with the helium bath. Left to itself, the sample will eventually cool to 4.2 K, but it may be maintained at some higher temperature by electrical heating. Consider the system shown in Figure 4.14. Thermal contact between the sample, which is in a vacuum, and the helium bath at 4.2 K is maintained by conduction along a thin walled stainless steel tube. The rate at which heat is conducted along the tube, length l and cross-sectional area A, is given by

$$\dot{Q} = \frac{A}{l}(T_s - 4.2)\,\bar{K},$$

where T_s is the temperature of the sample and $\bar{K}$ the integrated thermal conductivity of the tube between 4.2 K and T_s,

$$\bar{K} = \frac{\int_{4.2}^{T_s} K(T)\,dT}{T_s - 4.2}.$$

The sample is in thermal equilibrium when the electrical input to the heater P_e is equal to the heat conducted along the tube. Hence, by varying P_e, T_s may be controlled.

EXAMPLE 2

It is intended to use the system in Figure 4.14 to perform experiments on a sample at a temperature of 24 K. The sample is connected to the helium bath via a stainless steel tube which has a cross-sectional area of 10^{-6} m and a length of 2.5 cm. What electrical power input is required in order to maintain the sample at a temperature of 24 K?

The rate at which heat is conducted along the tube at a sample temperature of 24 K is

$$\dot{Q} = 4\times10^{-5}\int_{4.2}^{24} K(T)\,dT \text{ Watts};$$

using the data in Figure 2.6 gives $\dot{Q} = 0.96$ milliwatts, therefore

$$P_e = \dot{Q} = 0.96\,\text{mW}.$$

In most situations, there are stray sources of heat input to the sample region. It is unlikely that these would be constant with time, and so some form of temperature regulation is required. In its simplest form, this is provided by measuring the sample temperature with an electrical thermometer (see Chapter 7) and, by means of the appropriate electronics or a computer, comparing it with the required temperature. A signal, proportional to the temperature error, is used to increase or decrease the heater power as appropriate.

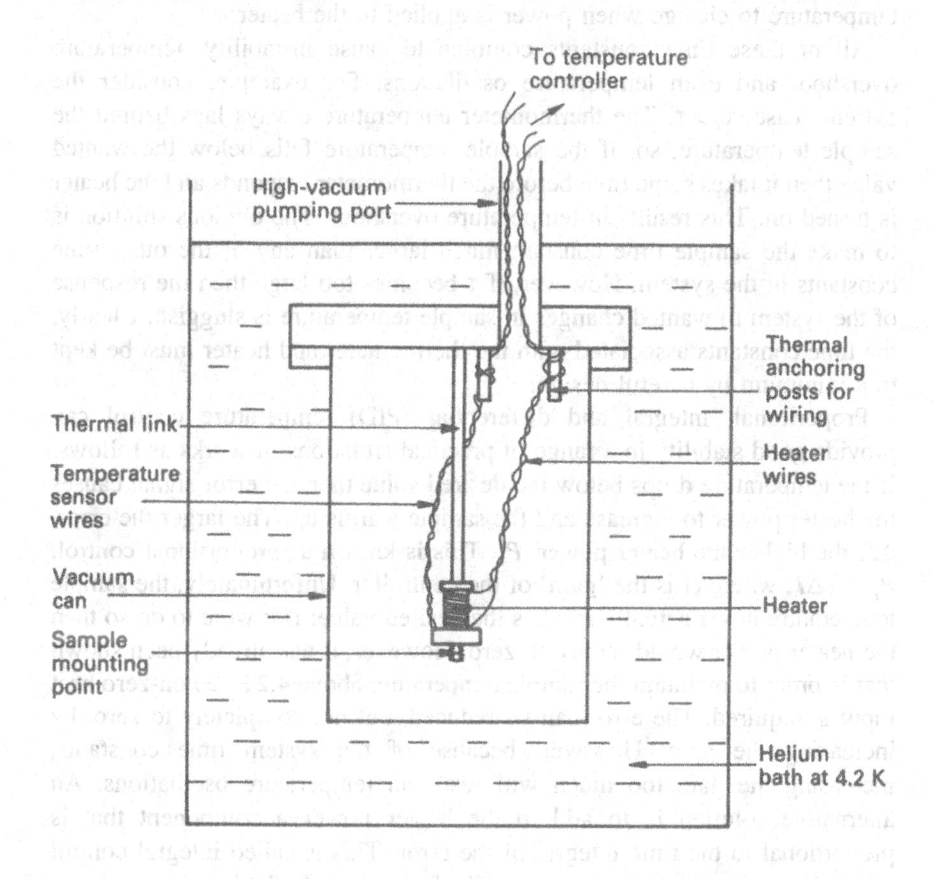

Figure 4.14 *Experimental arrangement for obtaining sample temperatures above 4.2 K*

In practice, it is difficult to maintain a high degree of temperature stability with such a simple system as this. The problem is that there are thermal time constants associated with the sample (and any mounting), the thermometer and the heater. The thermal time constant of the sample, τ, is a measure of the time taken for the sample to cool down via its thermal link, i.e. $\tau = RC$, where R is the thermal resistance of the link in K W^{-1} and C the heat capacity of the sample in J K^{-1}. Suppose that in the above example the sample is a 10-g block of copper. This has a heat capacity of about 0.15 J

K^{-1}. The thermal resistance of the link is $l/\bar{K}A = 2.1 \times 10^3\ K\ W^{-1}$. Therefore, the thermal time constant is $RC = 315$ s. The thermal time constant of the thermometer, τ_T, is a measure of the time taken for it to respond to a change in temperature of the sample. It depends on the thermal resistance of the contact between the sample and thermometer, the heat capacity of the thermometer and the response time of the electronics. The thermal time constant of the heater is a measure of the time taken for the sample temperature to change when power is applied to the heater.

All of these time constants combine to cause instability, temperature overshoot and even temperature oscillations. For example, consider the extreme case, $\tau_T > \tau$. The thermometer temperature always lags behind the sample temperature, so, if the sample temperature falls below the wanted value then it takes some time before the thermometer responds and the heater is turned on. This results in temperature overshoot. The obvious solution is to make the sample time constant much larger than any of the other time constants in the system. However, if τ becomes too large then the response of the system to wanted changes in sample temperature is sluggish. Clearly, the time constants associated with the thermometer and heater must be kept to a minimum by careful design.

Proportional, integral and differential (PID) temperature control can provide good stability in a range of practical situations. It works as follows: If the temperature drops below the desired value then the error signal causes the heater power to increase and the sample warms up. The larger the error, ΔT, the higher the heater power, P_e. This is known as proportional control, $P_e = G\,\Delta T$, where G is the 'gain' of the controller. Unfortunately, the sample temperature never actually reaches the desired value; if it were to do so then the heater power would reduce to zero. However, it has already been shown that in order to maintain the sample temperature above 4.2 K, a non-zero heat input is required. The error can be reduced, but not completely to zero, by increasing the gain. However, because of the system time constants, increasing the gain too much will result in temperature oscillations. An alternative solution is to add to the heater power a component that is proportional to the time integral of the error. This is called integral control and reduces the error to zero, upon which the input to the integrator is zero and its output becomes constant at the level required to maintain the desired sample temperature. However, the integrator also introduces another time constant into the system, which can increase the tendency towards instability or overshoot. Finally, to increase stability, a 'damping' term proportional to the rate of change of sample temperature can be added to the heater signal. This is the so-called derivative term. It is important to add the three contributions in exactly the right proportions, or the temperature accuracy will be worse than could be obtained by using proportional control on its own. The instructions provided with commercially available PID temperature controllers include detailed information on how to set them up for stable operation. A more detailed explanation of temperature control is given in the article by Forgan (1974), and is particularly helpful to those who wish to construct their own temperature control systems.

4.6 Obtaining temperatures below 4.2 K

Temperatures below 4.2 K are achieved by reducing the vapour pressure over a bath of helium. Figure 4.15 shows the temperature of a bath of ^{4}He as a function of its vapour pressure. With a sufficiently powerful pump, temperatures down to about 0.75 K should be theoretically possible. In practice, it is possible to obtain temperatures between 4.2 and about 1 K. Cooling to below the λ-point, 2.17 K, where liquid ^{4}He becomes superfluid, is difficult. Surfaces in contact with superfluid are covered with a helium film a few atoms thick (see Section 3.1.1). This flows rapidly until it reaches warmer parts of the apparatus and then evaporates, thus increasing the amount of gas the pump has to cope with. The film flow can be reduced by reducing the size of the orifice through which it must flow (the thickness of the film being constant), but this also introduces a restriction in the gas flow, so limiting the cooling power of the system.

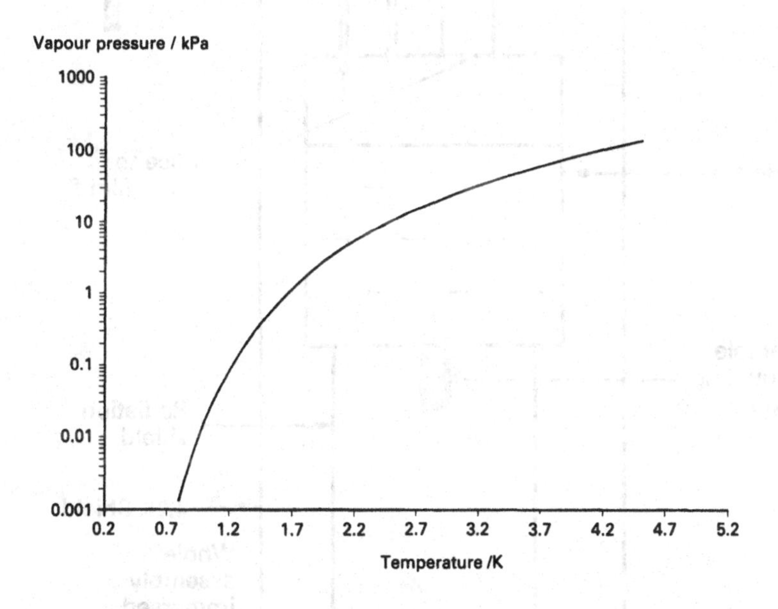

Figure 4.15 *Boiling temperature of liquid ^{4}He as a function of vapour pressure*

The cooling power of a pumped-liquid-helium system is given in terms of the molar latent heat of vaporisation, $L(T)$, and the molar evaporation rate, $\dot{n}$, in moles per unit time by

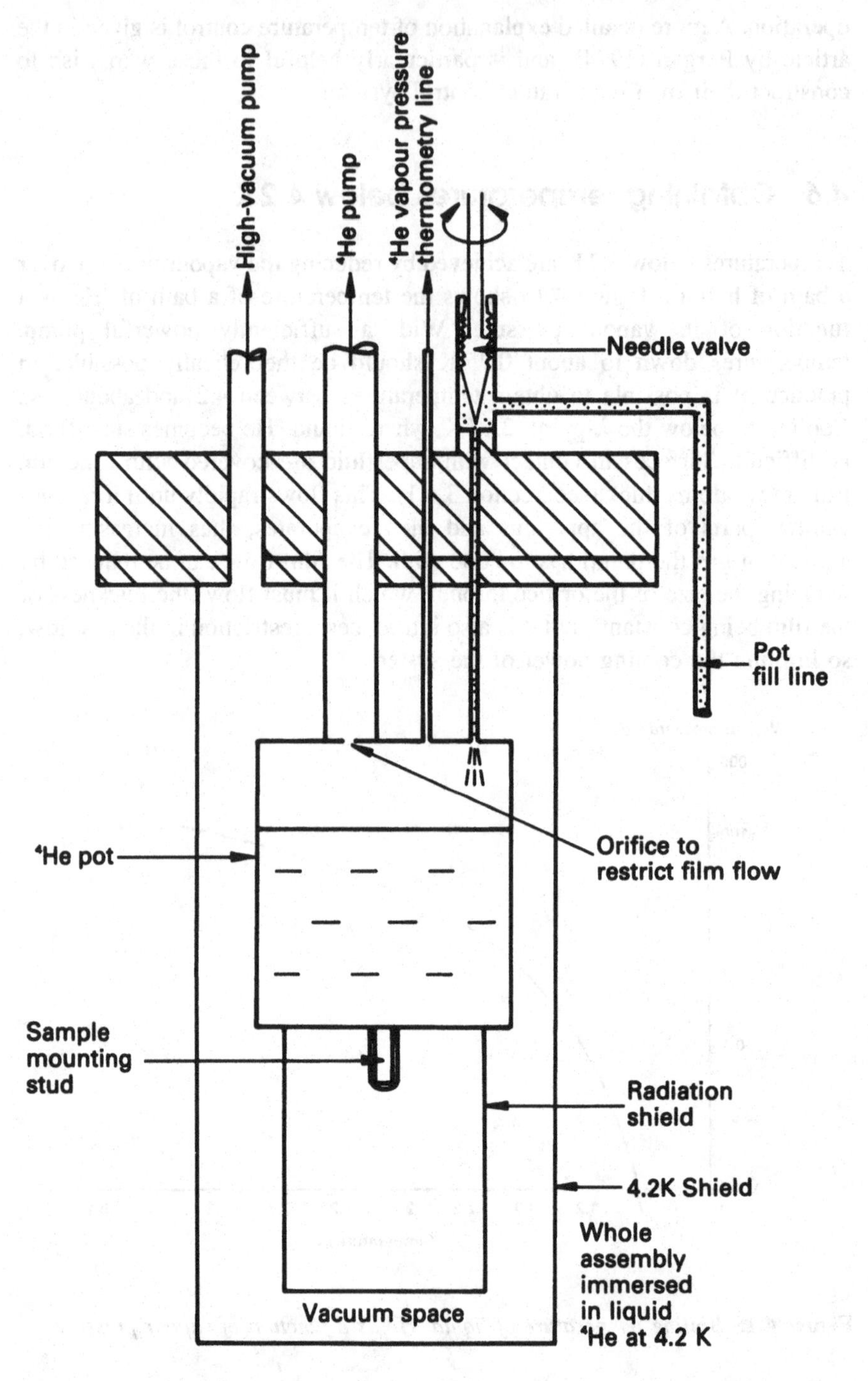

Figure 4.16 *System for obtaining temperatures between 4.2 K and about 1 K by pumping on a pot of liquid ^{4}He*

$$\dot{Q}(T) = L(T)\,\dot{n}.$$

The amount of liquid helium that must be evaporated, in order to cool the sample to the required temperature, depends on the total heat capacity of the sample plus liquid helium and container. It is often found that the dominant contribution to the total heat capacity is that from the liquid helium itself (at 4.2 K, 1 litre of liquid ^{4}He has a heat capacity of 400 J K^{-1} compared with only 0.2 J K^{-1} for a kilogram of copper). The pump may be attached to the main helium reservoir in which the sample is immersed. However, this proves rather wasteful, because about 40 per cent of the initial volume of helium must be evaporated just to cool the remaining volume from 4.2 to 1.5 K. Figure 4.16 shows a typical arrangement for cooling samples to ~1 K using a small ^{4}He pot that is thermally isolated from the main bath. It includes an arrangement to fill the pot from the main bath via a needle valve operated by a shaft from room temperature. The temperature is constant when the cooling power is equal to the heat leak to the helium pot. Clearly, the larger is the heat leak, the bigger is the pump that is required to maintain a given temperature.

EXAMPLE 3

What is the minimum pumping speed in litres per minute required to maintain a temperature of 1.2 K in the helium pot, if the total heat leak at this temperature is 10 mW? (The latent heat of vaporisation of ^{4}He in the temperature range 1.2 to 4 K is 96 J mol^{-1}.)

To maintain the temperature at 1.2 K, the number of moles of helium that have to be evaporated per second is given by

$$\dot{n} = 10^{-2}/96 = 10^{-4}\ \text{mol s}^{-1}.$$

The vapour pressure, p, corresponding to a temperature of 1.2 K is from Figure 4.15, 82 Pa (0.82 mbar). The vapour pressure is in equilibrium when the pumping speed equals the evaporation rate. If the pump is at room temperature, 300 K, then the (volume) pumping speed required is

$$\dot{V} = \frac{300\dot{n}R}{p} = 3.04 \times 10^{-3}\ \text{m}^3\,\text{s}^{-1} = 182\ l\,\text{min}^{-1}.$$

This is readily achieved by using a large rotary vacuum pump. To avoid vibration and noise problems, the pump should be mounted well away from the cryostat, preferably outside the room. It is important, therefore, to take account of the effect of the connecting pipes on the pumping speed at the cryostat (see Section 8.1).

The temperature is regulated by controlling the pressure in the helium reservoir. This is done most easily by manually operating a throttling valve in the pumping line. Better temperature stability can be achieved by using a manostat, an example of which is shown in Figure 4.17. A rubber diaphragm separates two chambers. In one chamber, the reference volume, is trapped a pressure corresponding to the required temperature; the other chamber is connected to the pumping line. The pressure in the reference volume is set by opening both the bypass and reference valves, which renders the manostat inoperative. When the pressure in the pumping line is at the required value, the two valves are closed. Now, if the pressure in the pumping line rises above the reference pressure then the diaphragm flexes and opens the pumping line. If, on the other hand, the pressure is too low then the diaphragm flexes back and closes off the pumping line. Other arrangements have been developed to regulate the pot temperature, e.g. by using electronic pressure or temperature sensors and electrically operated valves connected via a microcomputer.

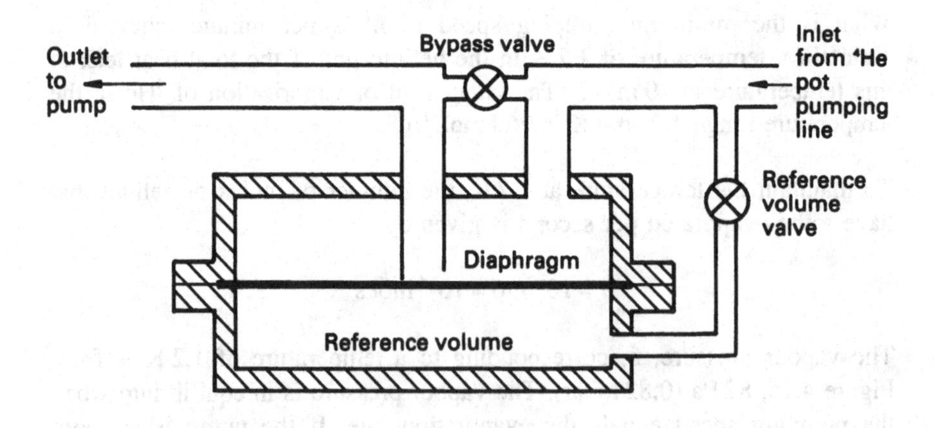

Figure 4.17 *Pressure regulator (manostat) used to maintain a constant temperature in a system like that shown in Figure 4.16*

4.7 Thermal contact and the Kapitza resistance

Unless the objective is to study the properties of the cryogen itself, the cold liquid must be used to reduce the temperature of a sample. This involves thermal contact between the two and perhaps an intermediary, like the pot containing the liquid. When any two materials are connected together, there is generally a thermal boundary resistance, R_B, which affects the transport of heat between them. If a heat flux $\dot{Q}$ passes across the boundary then a temperature difference, $\Delta T = \dot{Q}R_B$, appears across it. This situation is analogous to resistance in an electrical circuit, where the heat flux is like the electric current and the temperature difference is like the potential drop across the resistor. The thermal boundary resistance can have a considerable effect on the behaviour of low-temperature systems. When two metals are joined, the boundary resistance is minimised by ensuring good electronic contact. The surfaces must be clean and clamped tightly or soldered together. In the absence of any electronic contact, thermal conduction is due to phonons crossing the boundary. When phonons are incident on the interface between two materials having differing acoustic properties, not all are transmitted. The transmission coefficient depends on the phonon polarization, the angle of incidence on the interface and the degree of acoustic mismatch across the interface. Those phonons that are not transmitted are reflected back from the interface. This introduces a thermal boundary resistance, known as the Kapitza resistance, which cannot, like other causes of excess boundary resistance, be overcome by cleaning the surfaces and clamping carefully.

The Kapitza resistance is strongly temperature dependent, varying as T^{-n} where $n \approx 3$, so is much more serious at low temperatures. Typically, the thermal boundary resistance between a metal and an insulator is in the region of $0.5–1 \times 10^{-3}\,T^{-3}$ K m^2 W^{-1}. There is also a Kapitza resistance between liquid helium and solid surfaces, so if there is a heat input to a sample then it may be at a slightly elevated temperature, with respect to the helium bath in which it is immersed. Fortunately for the low-temperature experimenter, the liquid-helium–solid boundary resistance is significantly less than has been predicted, using the same theoretical models that have been proved to work well for solid–solid interfaces. This anomaly can be as large as one or two orders of magnitude. Considerable theoretical and experimental effort has been directed into finding the reasons for the anomaly (for reviews see Challis, 1974, and Wyatt, 1980). The cryostat designer is dependent on experimental measurements of the boundary resistance, which is found to be a strong function of the quality of the metal surfaces. For a smooth and clean copper surface, of area A m^2, in contact with helium-II, the Kapitza resistance is approximately $2 \times 10^{-3}/AT^3$ K W^{-1} (provided that the temperature difference across the boundary is not too large, i.e. $\Delta T \ll T$). This means that a copper sample of surface area 10 cm^2 and dissipating 10 mW, immersed in liquid ^4He at 1 K, will be at a temperature of about 1.02 K.

Significant advances in the understanding of the low-temperature properties of matter have been made using the relatively uncomplicated apparatus described in the preceding pages. It is now a fairly routine matter to perform experiments at temperatures down to 1 K, using the wide range of commercially built cryostats available. A number of experimenters will still prefer to build their own cryostats and they are referred to the book by A. C. Rose-Innes in the bibliography for more detailed design information.

Bibliography

Challis, L. J. (1974). *J. Phys. C (Solid State Phys.)*, **7**, 481

Forgan, E. M. (1974). *Cryogenics*, **14**, 207

Wyatt, A. F. G. (1980). In *Phonon Scattering in Condensed Matter*, edited by H. J. Maris (New York: Plenum Press)

The standard work on experimental techniques at low temperatures, as relevant today as when it was written, and a must for anyone embarking on experimental work at low temperatures:

Rose-Innes, A. C. (1973). *Low Temperature Laboratory Techniques* (London: English Universities Press)

5

Reaching low temperatures, stage 2: ^{3}He and ^{3}He-^{4}He cryogenic systems 1 K-1 mK

For the reasons already discussed in Chapter 4, it is very difficult to obtain temperatures much below 1 K by pumping on liquid ^{4}He. The size and complexity of the necessary pumping system increases rapidly for every fraction of a degree lower that is wanted. If lower temperatures are needed, they may be obtained using the lighter isotope ^{3}He.

5.1 The ^{3}He cryostat

Temperatures down to about 0.3 K may be obtained by pumping on a bath of ^{3}He. The vapour pressure at a given temperature (see Figure 5.1a) is higher in ^{3}He than in ^{4}He and there is no superfluid film flow to contend with. However, ^{3}He is very expensive. It is produced by the radioactive (β^-) decay of tritium, which has a half-life of 12.3 years. Consequently, ^{3}He cryogenic systems are designed to work with just a few cubic centimetres of liquid. An example of a pumped ^{3}He cryostat is shown in Figure 5.1b. Continuous cooling in the pot can be maintained by recirculating the ^{3}He. The sample is mounted in vacuum, in thermal contact with the ^{3}He pot. The pot and sample are surrounded by a radiation shield, which is held at about 1.3 K by a pumped ^{4}He pot. Initially, the sample is cooled to 1.3 K by allowing low-pressure ^{4}He exchange gas into the vacuum space. The gas conducts heat between the sample and the 1.3 K shield. After pumping out the exchange gas, the ^{3}He gas is allowed into the cryostat. It passes through the condenser tube at 1.3 K and enters the pot as liquid, where its temperature may be reduced by pumping. To increase the pumping speed at low pressures, a vapour diffusion pump may be used in conjunction with the rotary pump (see Section 8.1). The pumps and associated pipework are carefully sealed to avoid loss or contamination of the precious gas. After passage through a purification trap, the ^{3}He is returned to the pot via the condenser. The flow restriction between the condenser and the pot is to ensure that sufficient pressure is built up in the condenser for liquid to form. Continuous-flow systems like this are not able to attain the lowest temperatures. This is because heat is added to the pot via the condensed liquid. Some arrangement may be included to stop the return flow and pump on the pot flat out, in which case the ^{3}He gas is collected in storage canisters for re-use. With a typical pumping system having a maximum speed, at the cryostat, of 7500 l (at 300 K) per minute at a pressure of 2.5×10^{-3} mbar, it is possible to achieve a cooling power of 0.1 mW at a temperature of 0.3 K using a cryostat such as this. The temperature may be controlled simply by regulating the pumping speed (see Section 4.6).

At the very low pressures corresponding to temperatures below about 0.3 K, it is hard to obtain a sufficiently high pumping speed when using the conventional mechanical–diffusion pump arrangement. Temperatures down to about 0.25 K can be achieved by using a charcoal sorption pump instead. The sorption pump works on the cryopump principle. Activated charcoal, having a large surface area for small volume, is cooled to liquid helium temperatures. It is then able to absorb helium gas to the extent of 0.1 grams

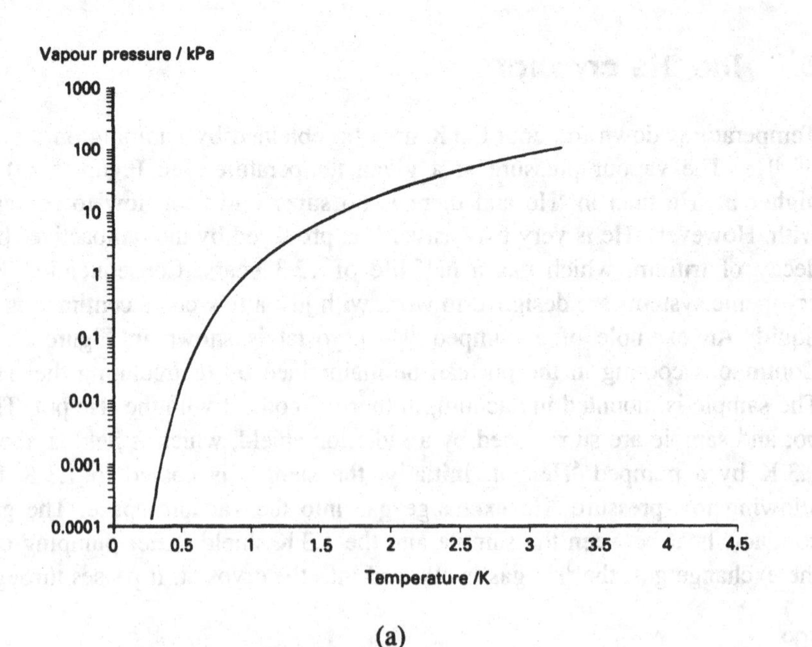

(a)

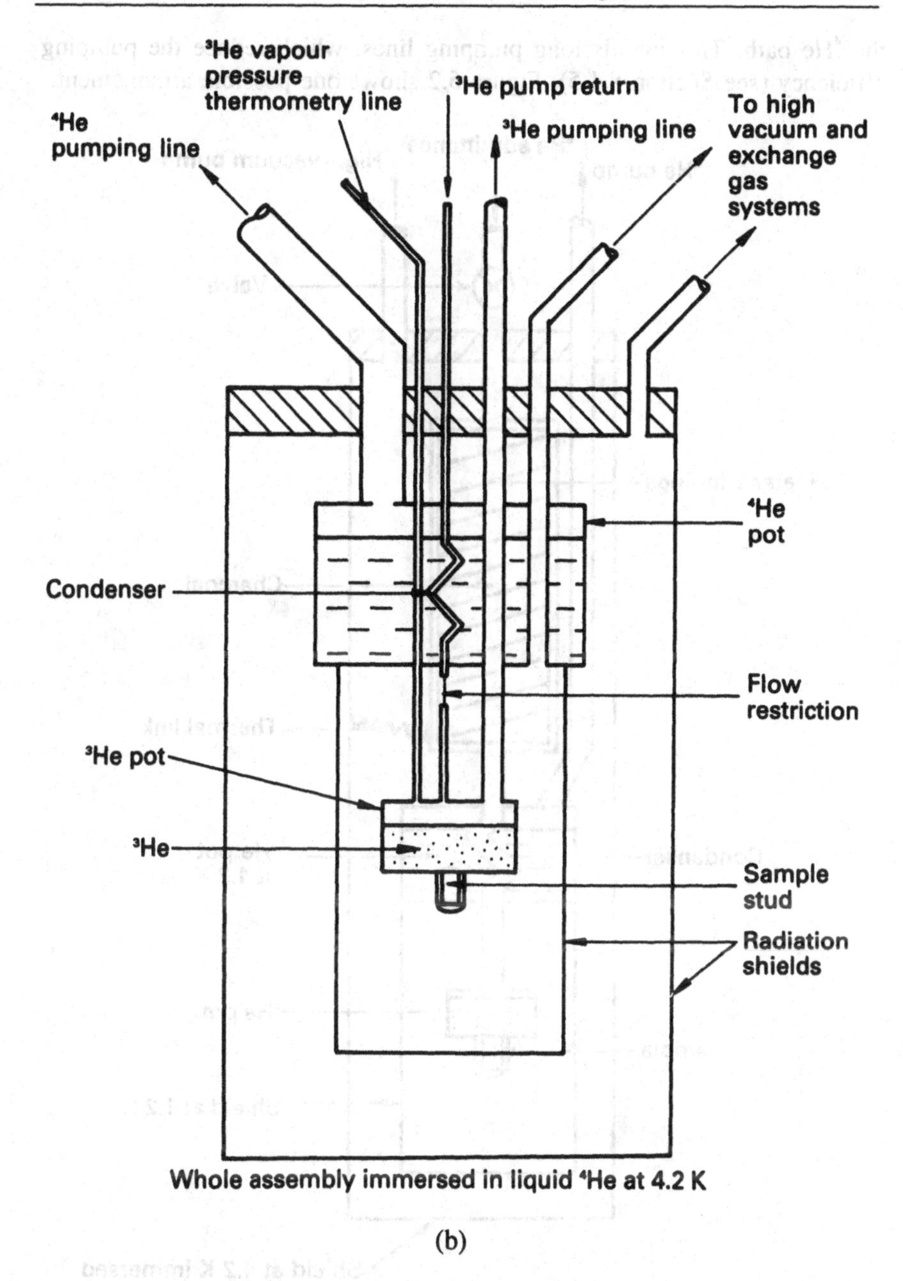

(b)

Figure 5.1 (a: below, left) *Vapour pressure of ^{3}He as a function of temperature.* (b: above) *Continuous-cycle ^{3}He refrigerator used for obtaining temperatures down to approximately 300 mK*

of gas per gram of charcoal, independent of gas pressure. An advantage of the sorption pump is that it can be located inside the cryostat and cooled by the ^{4}He bath. This avoids long pumping lines, which reduce the pumping efficiency (see Section 8.1.5). Figure 5.2 shows one possible arrangement.

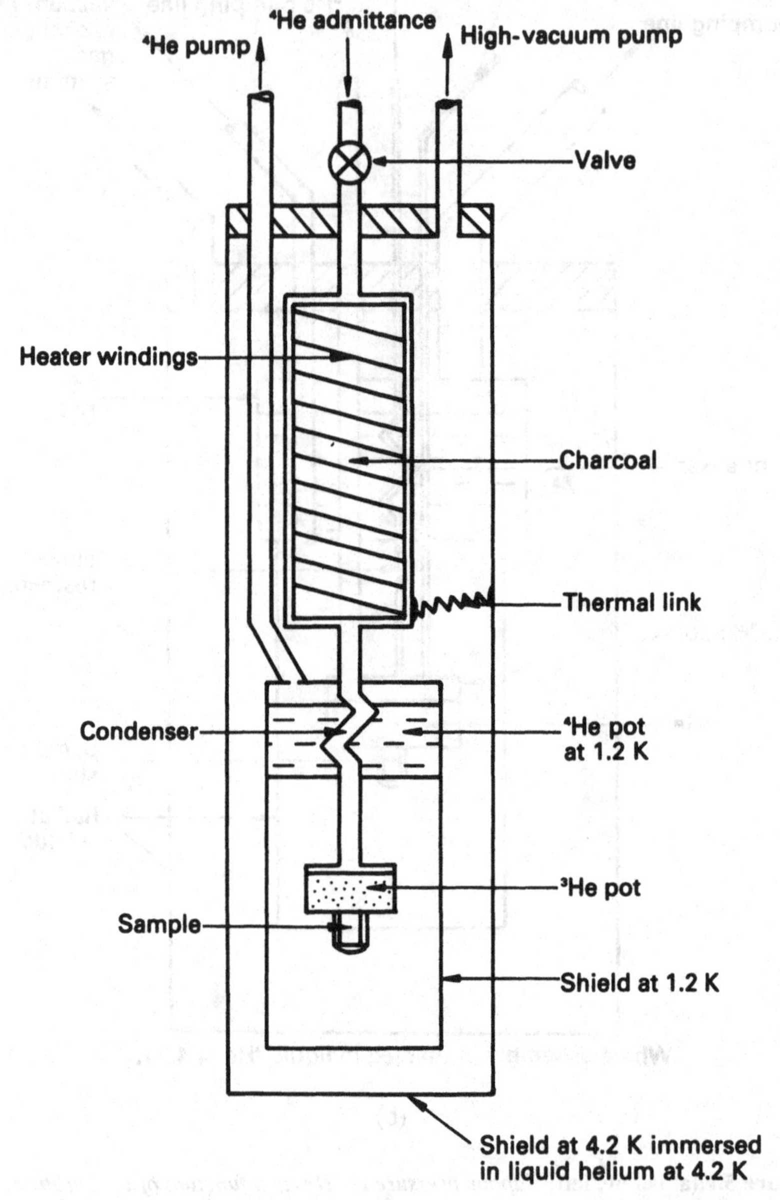

Figure 5.2 *Sorption pumped ^{3}He cryostat, see the text for an explanation of its operation*

The canister containing charcoal is in weak thermal contact with the ^{4}He bath. ^{3}He gas is admitted to the top of the cryostat and is immediately absorbed by the charcoal at 4.2 K. The inlet valve is closed, and when the charcoal is heated to around 40 K, the helium is desorbed and passes through the condenser, where it liquefies before entering the ^{3}He pot. When the charcoal is allowed to cool slowly it starts to absorb the vapour and the pot temperature reduces. The disadvantage of this process is that it is 'single-shot'; however, it may be repeated by reheating the charcoal, etc.

5.2 The ^{3}He–^{4}He dilution refrigerator

The dilution refrigerator is the workhorse of the millikelvin temperature range. First proposed by London, Clarke and Mendoza in 1962, the dilution refrigerator takes advantage of the peculiar low-temperature behaviour of the ^{3}He–^{4}He mixtures (see Section 3.3). For concentrations of ^{3}He greater than about 6.5 per cent, phase separation into a ^{3}He-rich, concentrated phase floating on top of a dilute, ^{4}He-rich phase takes place. The dilute phase is so called because, as the temperature approaches absolute zero, there is a maximum solubility of 6.5 per cent ^{3}He in ^{4}He (Figure 3.15). The superfluid ^{4}He just acts as an inert background for a dilute Fermi gas of ^{3}He. This implies that if one takes pure superfluid ^{3}He below 0.86 K and starts adding ^{3}He, then it is energetically favoured for the first ^{3}He atoms to go straight into solution, but when the concentration reaches 6.5 per cent the ^{3}He prefers to form a concentrated phase. If it were somehow possible to take ^{3}He atoms out of the dilute phase, then to maintain the equilibrium concentration they would be replaced by atoms crossing the boundary from the concentrated phase. This is analogous to evaporation from the concentrated phase which, consequently, cools as the 'latent heat of vaporisation' is removed, and is the underlying principle of operation of the dilution refrigerator. Continuous cooling by this method is possible because, unlike cooling by evaporation (see Section 4.5), where the vapour pressure reduces rapidly with temperature, making it increasingly difficult to maintain a reasonable cooling power, the 'vapour' or dilute phase concentration is constant at about 6.5 per cent.

When a mole of ^{3}He reversibly crosses the boundary from the concentrated phase to the dilute phase, the heat absorbed from the surroundings depends on the enthalpy difference between the concentrated and dilute phases of ^{3}He

$$\Delta Q = \Delta H = T\,\Delta S,$$

where T is the temperature of the mixture. Hence

$$\Delta Q = T(S_d(T) - S_c(T)),$$

where $S_d(T)$ and $S_c(T)$ are the molar entropies of ^{3}He in the dilute and concentrated phases, respectively. Therefore, the cooling power resulting from $\dot{n}_3$ moles of ^{3}He crossing the phase boundary per second is

$$\dot{Q} = \dot{n}_3 T(S_d(T) - S_c(T)).$$

There is no volume change in this process; application of the first law of thermodynamics (see Section 1.2.2) gives

$$dQ = dE = c_V dT.$$

Therefore,

$$T\,dS = c_V\,dT.$$

Integrating this gives

$$S = \int \frac{c_V}{T}\,dT.$$

Below about 100 mK, ^{3}He is assumed to be a degenerate Fermi gas. The specific heat capacity per mole of an ideal degenerate Fermi gas is given by (see Section 2.3.2)

$$c = \frac{\pi^2 N k^2 T}{2E_F}.$$

Hence

$$S = \frac{\pi^2 k^2 N m^*}{\hbar^2}\left(\frac{V}{3\pi^2 N}\right)^{2/3} T,$$

where N is Avogadro's number and V the molar volume of ^{3}He. Substitution of the relevant data for ^{3}He in the concentrated and dilute phases into this expression (see Table 5.1) yields $S_c = 23T$ and $S_d = 107T$. The theoretical cooling power is therefore

$$\dot{Q} = \dot{n}_3(107 - 24)T^2 = 84\dot{n}_3 T^2 \text{ W} \quad (\text{for } T < 100 \text{ mK}).$$

Table 5.1 Properties of ^{3}He in the concentrated and dilute phases

	Concentrated phase	Dilute phase
Effective mass, m^*	$2.8m_3 = 1.45 \times 10^{-26}$ kg	$2.4m_3 = 1.20 \times 10^{-26}$ kg
Molar volume, V	3.7×10^{-5} m^3	4.6×10^{-4} m^3

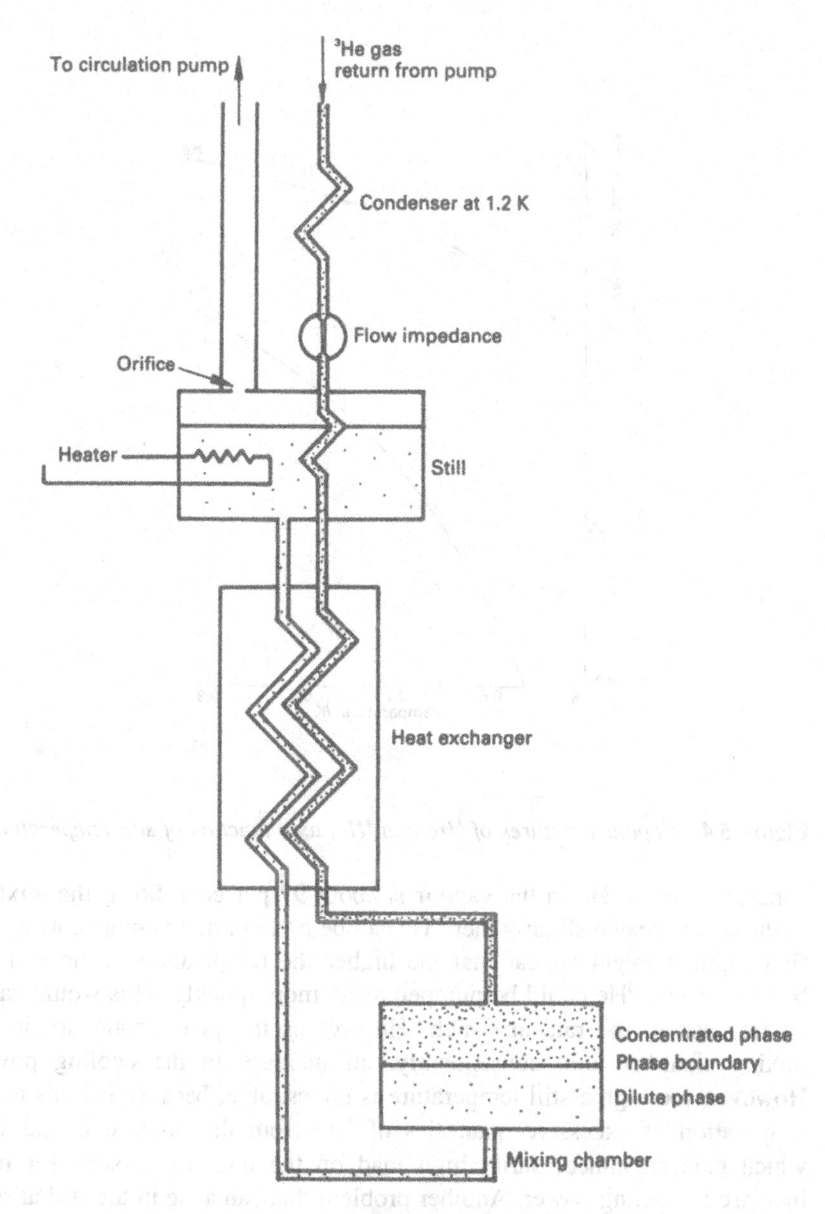

Figure 5.3 *Schematic diagram of a ^3He–^4He dilution refrigerator*

In a practical dilution refrigerator (Figure 5.3), cooling takes place in the mixing chamber. The flow of ^3He across the boundary between the phases is maintained by distilling the ^3He off from the dilute phase. This is achieved in a *still* that is connected to the bottom of the mixing chamber. Below about 1 K the vapour in equilibrium with a dilute ^3He–^4He mixture is composed mainly of the more volatile ^3He (see Figure 5.4). For a mixture at 0.7 K, the

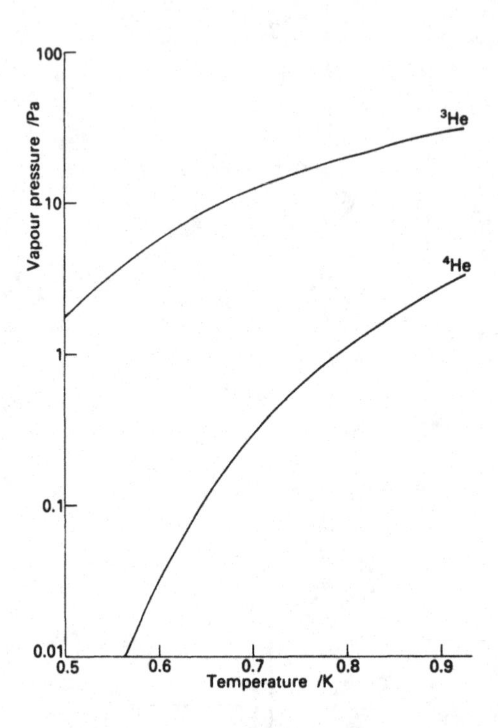

Figure 5.4 *Vapour pressures of ^{3}He and ^{4}He, as a function of still temperature*

concentration of ^{3}He in the vapour is about 97 per cent. So, if the mixture in the still is heated slightly then ^{3}He can be preferentially pumped away. At first sight, it might appear that the higher the temperature in the still the better, because ^{3}He could be pumped away more quickly. This would cause an increase in the rate at which ^{3}He crosses the phase boundary in the mixing chamber and, consequently, an increase in the cooling power. However, too high a still temperature is undesirable, because it leads to the evaporation of excessive quantities of ^{4}He from the mixture in the still, which puts an unnecessarily high load on the pumping system for little increase in cooling power. Another problem that can arise in the still at such low temperatures is superfluid ^{4}He film flow. This causes ^{4}He to creep up into the pumping tube in which it may reach regions where the temperature is sufficiently high to evaporate it, thus putting extra strain on the pumps. Careful still design can partially overcome this problem. For example, placing a small orifice in the pump outlet will considerably reduce the film flow. Typical still temperatures are in the region 0.6–0.7 K, requiring an electrical input, P_e, to the still heater given by

$$P_e = \dot{n}_3(H_v(T) - H_l(T)),$$

where $H_l(T)$ and $H_v(T)$ are respectively the enthalpies of the liquid and the vapour phases at the still temperature T. This yields an electrical power requirement of approximately $40\dot{n}_3$ watts, which may be reduced if there are heat leaks from higher-temperature parts of the apparatus.

To maintain continuous cooling, ^{3}He removed from the still is recirculated, via the pump, to the mixing chamber (see Figure 5.6). After being cleaned in a charcoal trap at 77 K and re-entering the cryostat, the ^{3}He gas is condensed in a line running through the ^{4}He pot at 1.3 K. A flow restriction is positioned after the condenser, to ensure that sufficient pressure is built up to condense the gas. It is usually fabricated from a very fine stainless steel capillary tube. To restrict the flow still further, a fine wire may be inserted into the tube. The condensed ^{3}He is then returned to the mixing chamber via a set of heat exchangers. In the heat exchangers, the returning liquid is cooled by the dilute-phase ^{3}He flowing between the mixing chamber and the still. This ensures that the ^{3}He is returned to the mixing chamber at near to the same temperature as the liquid already there, thus minimising unwanted heat input and the base temperature that can be reached.

The heat exchangers should, ideally, possess the same qualities as those used in helium liquefiers. However, fabricating a heat exchanger with a high thermodynamic efficiency is more difficult at lower temperatures. For example, the Kapitza resistance is more serious at millikelvin temperatures. This makes the requirement for a large contact area between the liquid and the heat exchanger paramount. Also of importance is minimising the volume of the exchanger to reduce the amount of expensive ^{3}He required to operate the refrigerator. Sintered-metal heat exchangers (Figure 5.5) are best able to satisfy these requirements. Small, 50-μm particles of copper or silver are compressed together to form a 'sponge' through which the liquid passes. Two such sponges, one for the incoming liquid and another for the outgoing liquid, are thermally linked via many small wires of the same metal embedded in the sponge (a similar constructional technique may be employed in the mixing chamber to maximise heat transfer between the helium mixture and the metal chamber on which the sample is mounted). A number of these step exchangers, so called because each one operates at a constant temperature, are connected in series via a conventional counterflow type heat exchanger.

It is important to minimise the amount of ^{4}He in the recirculated gas, not just to reduce load on the pumping system as explained earlier, but also to minimise the heat input to the mixing chamber. As the recycled mixture is condensed and subsequently cooled, any ^{4}He present will separate out and form a dilute phase. Further cooling of the mixture in the heat exchangers leads to a condensation of ^{3}He from the dilute to the concentrated phase, the opposite of the process taking place in the mixing chamber, and one which liberates heat. Furthermore, the superfluid dilute phase has a high thermal conductivity and brings heat directly from the higher-temperature parts of the system to the mixing chamber.

A well-designed pumping system having a high throughput is needed in order to get the best performance out of a dilution refrigerator.

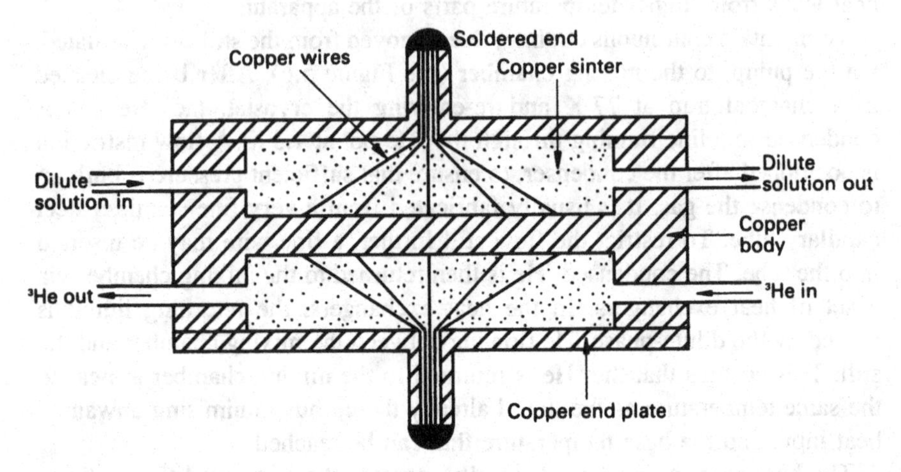

Figure 5.5 *Sintered metal step heat exchanger unit*

■ EXAMPLE

What is the minimum speed, in litres of gas per minute (at 300 K), of the pumping system required for use with a dilution refrigerator having a cooling power of 10 μW at a temperature of 10 mK? The still temperature is 0.7 K.

The number of moles of ^{3}He that have to be circulated per second is

$$\dot{n}_3 = 10^{-5}/84(10^{-3})^2 = 0.0012.$$

Assuming ^{3}He gas to be ideal, the volume throughput of the pump is given by

$$\dot{V} = \dot{n}_3 RT/p.$$

The vapour pressure of ^{3}He in the still at 0.7 K is, from Figure 5.4, 0.082 mbar or 8.28 N m^{-2}. Therefore, at 300 K,

$$\dot{V} = 0.0012 \times 8.31 \times 300/8.28 = 0.363 \text{ m}^3 \text{ s}^{-1} = 21\,800 \text{ l min}^{-1}.$$

This is quite a high pumping speed by most standards. To achieve it at such low pressures would require the use of a high-speed booster pump ahead of a large rotary pump (see Section 8.1) and large-diameter pumping tubes. In reality, the pumping speed will have to be even greater to account for the effect of the interconnecting pipes and any ^{4}He gas that is evaporated from the still.

In practice it is found that there is a lower limit to the base temperature that can be achieved by simply increasing the pumping speed. This is because, at high ^{3}He throughputs, viscous heating of the dilute phase leaving the mixing chamber becomes important. The heat dissipated in this process is conducted back to the mixing chamber and increases the heat loading.

It is normal to run a dilution refrigerator 'flat out' at its maximum cooling power. To obtain temperatures above the base temperature, electrical heating is applied to the mixing chamber. The mixing chamber temperature is proportional to the square root of the heater power, which is proportional to the heater current, I_h. Therefore, $T \propto I_h$. A thermometer mounted on the mixing chamber feeds back the temperature to a PID temperature controller (see Section 4.5), which continually adjusts the heater current to keep the temperature stable. To run at temperatures above about 100 mK, the still temperature can be reduced and the pumps throttled back. However, to achieve high-temperature stability, the electrical regulation system is still required.

Dilution refrigerators can be bought from a number of cryogenic equipment manufacturers. Alternatively, some experimenters opt to build their own small refrigerators. Figure 5.6 shows a cryostat incorporating a dilution refrigerator. With careful design to minimise stray heat leaks, such a system is capable of achieving temperatures down to about 10 mK with a net cooling power of the order of microwatts. To date, the lowest temperature achieved using a dilution refrigerator is in the region of 1–2 mK. A number of laboratories now possess the equipment to attain temperatures down to a few tens of millikelvins. A full review of dilution cooling can be found in the book by Lounasmaa listed in the bibliography at the end of this chapter.

5.3 Pomerantchuk cooling

Pomerantchuk cooling, or adiabatic compression of ^{3}He along the melting curve, may be used to achieve temperatures down to approximately 2 mK (Anufriev, 1965). It is particularly useful in experiments where the ^{3}He melting transition is being studied. To see how this technique works consider the p–T phase diagram of ^{3}He (Figure 5.7). At 319 mK there is a minimum in the melting curve a feature that is not seen in the phase diagram of ^{4}He (Figure 3.1). Applying the Clausius–Clapeyron equation,

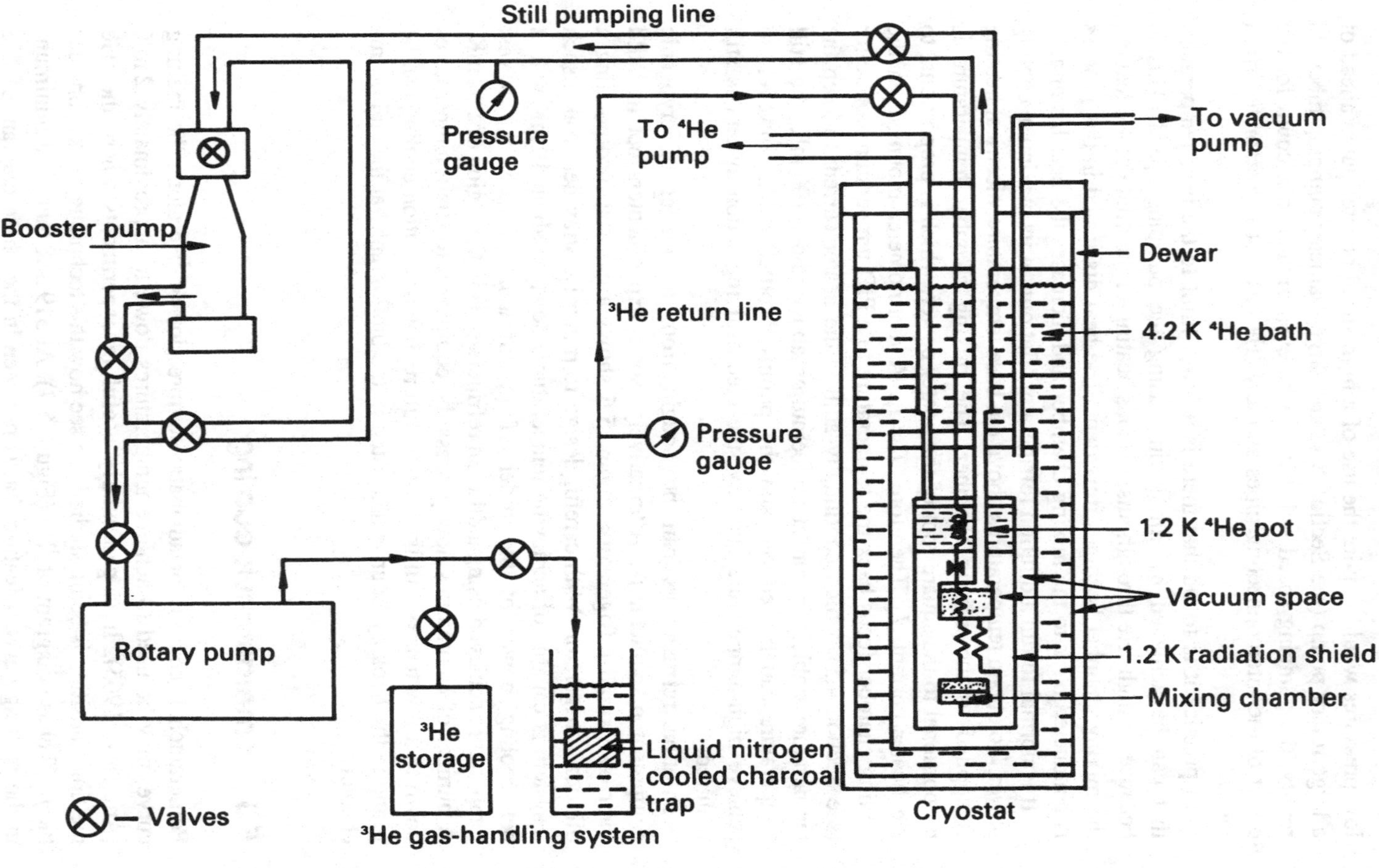

Figure 5.6 *Dilution cooling cryostat and its pumping system*

$$\frac{dP}{dT} = \frac{S_l - S_s}{V_l - V_s},$$

at the phase transition, where S_l, V_l and S_s, V_s are the entropy and volume of the liquid and solid phases respectively, implies that below 319 mK, because the volume increases on melting, the entropy of the solid is actually higher than that of the liquid! Why is this? To see why, it is necessary to consider the *total* entropy of the system, including the disorder in the arrangement of the magnetic dipoles or spins associated with the ^3He atoms. In the solid at low-temperatures, the entropy is due primarily to the disorder in the magnetic dipoles which are all oriented randomly. At temperatures above 1 mK the molar entropy of the solid is nearly independent of temperature and has a value $R \ln 2$. Around 1 mK, magnetic ordering of the dipoles takes place as the magnetic dipolar interactions begin to dominate over the thermal motion. This results in a sudden decrease in the entropy of the solid.

The entropy of the liquid may be expressed in terms of its specific heat capacity

$$S_l = \int \frac{c}{T} \, dT,$$

at low temperatures $c \propto T$ (see Figure 3.10) hence $S_l \propto T$. So, as the temperature is reduced towards absolute zero, the entropy of the liquid decreases with an approximately linear dependence on T, while the entropy of the solid is constant (Figure 5.8). Below 319 mK the entropy of the liquid is less than that of the solid giving rise the negative slope in the melting curve. Now, if the temperature and pressure were such that the ^3He was initially liquid and at a lower temperature than the minimum in the melting curve, point A in Figure 5.8, adiabatic compression to point B would solidify the ^3He and leave it at a lower temperature.

In practice, the initial conditions are set up at constant temperature, with the ^3He pressure cell in thermal contact with the mixing chamber of a dilution refrigerator. Then the cell is thermally isolated from the refrigerator and the ^3He compressed. A change in temperature of about one order of magnitude can be obtained with this method. The cooling power of a Pomerantchuk refrigerator is given by

$$\dot{Q} = \dot{n}_3 T (S_{sm} - S_{lm}),$$

where $\dot{n}_3$ is the number of moles of ^3He frozen per unit time, and S_{sm} and S_{lm} are respectively the entropies of the solid and liquid phases on the melting curve. As the base temperature of around 1 mK is approached, the decrease

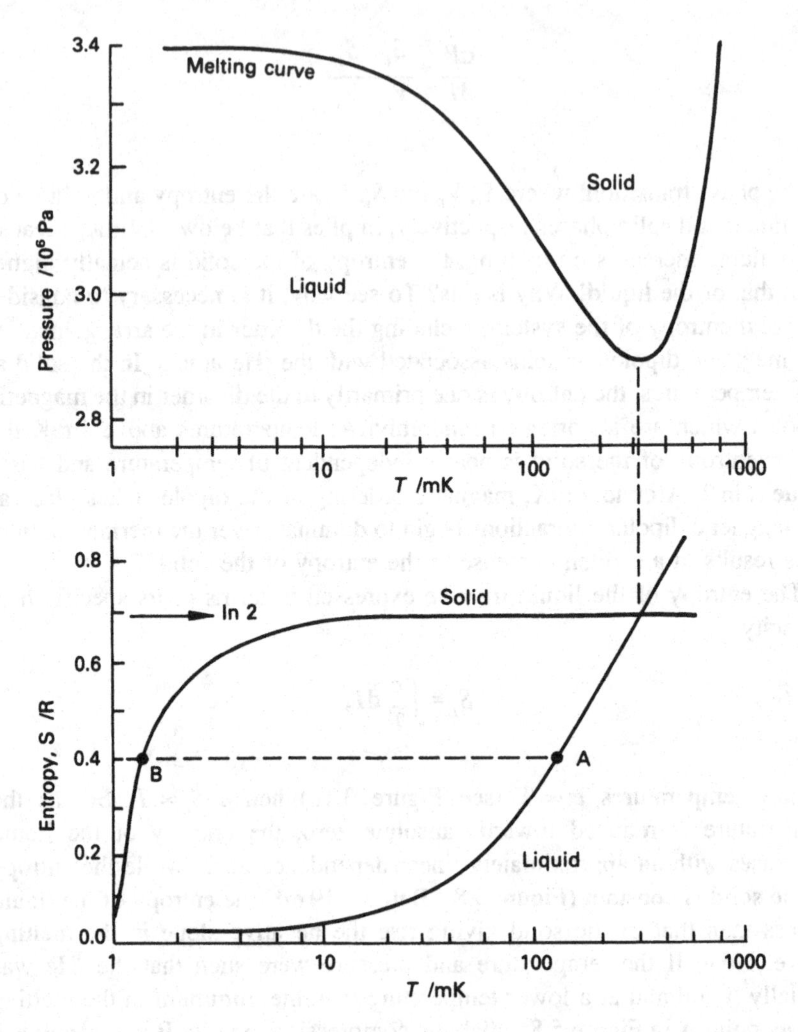

Figure 5.7 (top) *The ^{3}He melting curve* and **Figure 5.8** (bottom) *Entropy of ^{3}He in the liquid and solid phases (Betts, 1989)*

in entropy of the solid as magnetic ordering takes place leads to a decrease in the cooling power. Pomerantchuk cooling is a 'single-shot' process in that when all of the ^{3}He has been converted to solid it is necessary to warm the cell up and start again.

Figure 5.9 shows an experimental arrangement for Pomerantchuk cooling. Pressure is hydraulically transmitted from room temperature to the ^{3}He cell using ^{4}He. The ^{4}He operates flexible bellows which are linked to a second set of bellows containing the ^{3}He. Because ^{4}He has a lower freezing pressure than ^{3}He, a pressure differential of a few bar must be maintained between the two bellows. This is achieved by making the ^{4}He bellows have a larger area.

More sophisticated systems also have thermometry built into the ^{3}He pressure cell.

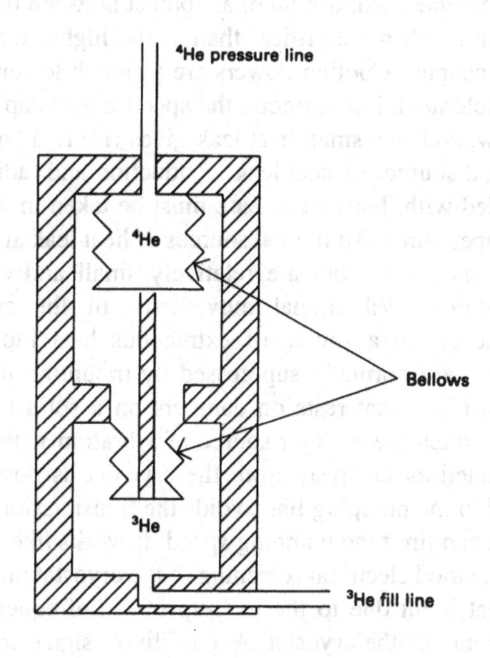

Figure 5.9 *Hydraulically operated Pomerantchuk refrigeration cell*

At low temperatures the Pomerantchuk refrigerator is capable of a much higher cooling power than a typical dilution refrigerator (at 10 mK, about an order of magnitude more per mole of ^{3}He solidified, than per mole circulated in a dilution refrigerator). It is also possible to achieve lower base temperatures in a Pomerantchuk cell, typically about 1 mK compared with 2 mK for a very good dilution refrigerator. However, Pomerantchuk refrigerators are very difficult to build and so are not widely used in modern laboratories. In the past, before efficient dilution refrigerators became commercially available, Pomerantchuk cooling enabled the first studies of ^{3}He along the melting curve to be undertaken. The first observation of the superfluid phases of ^{3}He at high pressure was made in a Pomerantchuk cell.

5.4 Working at millikelvin temperatures

The ultimate base temperature in continuous and one-shot cooling processes is determined by the balance between the refrigeration system's cooling power and the heat input due to stray heat leaks. Unwanted heat leaks also reduce the low-temperature hold time in single-shot cooling systems. When working at millikelvin temperatures, the precautions necessary to minimise heat leaks and maximise thermal contact between the sample and the cooling stage are much more critical than at the higher temperatures considered in the last chapter. Cooling powers are so much lower that much less heat leak can be tolerated. Furthermore, the specific heat capacities of the samples are very low, and so a small heat leak gives rise to a large temperature increase. The usual sources of heat leak, conduction and radiation, always have to be contended with, but greater care must be taken in dealing with them at very low temperatures. Additional sources of heat leak also become apparent (they are always present but are relatively small and can be ignored at higher temperatures). Vibrational movements of the cryostat and its internal components are a source of extraneous heat input. Externally generated vibrations are normally suppressed by mounting the cryostat on a massive concrete block that rests on dampers on a solid (preferably ground) floor. Pumps, which are a major source of vibration within the laboratory, should be mounted as far away from the cryostat as possible. A ballast chamber inserted in the pumping line avoids the transmission of pulses in the pumped gas, but can limit the pumping speed. It works like the silencers on a vehicle exhaust. Good electrical screening of the cryostat minimises the unexpectedly high heat input due to the pick-up of radiofrequency signals by the metal components of the cryostat. A 1 millivolt signal induced in a 1-metre-long stainless steel tube could lead to a power dissipation of 1 microwatt, which is comparable to the cooling power of a typical dilution refrigerator below 10 mK. For this reason, millikelvin cryostats are often constructed within large, metal-walled rooms. The walls are electrically earthed and form a screening cage around the cryostat. Another use of the screened room is to protect sensitive measuring instruments from interfering electromagnetic fields.

In spite of all the difficulties involved, a wide variety of experiments are carried out at temperatures down to about 1 mK. The properties of superfluid ^{3}He, ^{3}He–^{4}He mixtures and electrons in semiconductors are currently receiving a large amount of attention.

Bibliography

Anufriev, Y. D. (1965). *Soviet Physics–JETP*, **1**, 155

Betts, D. S. (1989). *An Introduction to Millikelvin Technology* (Cambridge: CUP)

London, H., Clarke, G. R. and Mendoza, E. (1962). *Phys. Rev.*, **128**, 1992

Lounasmaa, O. V. (1974). *Experimental Principles and Methods Below 1 K* (London: Academic Press)

6 Reaching low temperatures, stage 3: adiabatic demagnetisation, 1 mK-

Although the absolute zero of temperature can never be reached, as discussed in Chapter 1, it may be approached indefinitely closely. To date, the lowest temperature obtained in practice is of the order 10^{-8} K. This has been achieved by adiabatic demagnetisation of copper nuclei.

Adiabatic demagnetisation of electronic spins in certain paramagnetic salts, like cerium magnesium nitrate (CMN), has been used to achieve temperatures in the millikelvin range on a single-shot basis. The technique was suggested by Debye in 1925 and first demonstrated in 1933 by Giauque and MacDougall. Magnetic cooling pre-dated dilution cooling by 30 years and, for a long time, it was the only method of obtaining sub-1 K temperatures. In recent years it has been superseded by continuous-dilution cooling in the millikelvin range of temperatures. However, adiabatic demagnetisation of nuclear 'spins' in metals is currently the only method of obtaining microkelvin temperatures in large samples. The following discussions will be concerned primarily with nuclear demagnetisation, although the principles apply equally well to electronic demagnetisation.

6.1 Magnetic properties of matter at low temperatures

All materials exhibit magnetic behaviour. In the case of iron and other ferromagnets this is well known and easily demonstrated but in most materials the magnetic effects are rather weak. The magnetism arises from the particular microscopic nature of the atom. The motion of constituent charged particles in orbits or spinning about their axis can be visualised as tiny circulating currents which may give rise to small magnetic dipole moments. In many molecules and atoms, the circulating currents are balanced out and there is no net magnetic moment. The influence of an external field can unbalance the currents and induce a magnetic moment in the atom. The induced moment acts to oppose the applied field and in diamagnetic materials this is the dominant effect. In certain materials, those said to be paramagnetic, there are permanent magnetic dipoles arising from the net circulating current associated with each atom. The material can be visualised as an assembly of microscopic bar magnets.

In the absence of any external magnetic field the dipoles are all oriented randomly, owing to thermal motions of the atoms and molecules. Hence, a macroscopic sample of the material has no net magnetic moment. If an external field is applied, the dipoles experience a force which tends to line them up in the same direction as the applied field, outbalancing any natural diamagnetism and giving the sample a net positive magnetic moment. As the temperature is reduced, the thermal motions responsible for randomising the orientations of the magnetic dipoles become weaker, and the sample is found easier to magnetise. At still lower temperatures, as T approaches absolute zero, the small forces exerted by neighbouring dipoles on each other are sufficient to order them, even in the absence of an external magnetic field. It is found that metals which are not normally ferromagnetic can become so at very low temperatures (it is interesting to note that with few exceptions, those that do not, become superconducting instead). Many other paramagnetic systems make the transition to an antiferromagnetic state (Bleaney and Bleaney, 1976). Paramagnetic systems will now be singled out for more detailed consideration, as they are used in magnetic cooling.

Sub-atomic particles, electrons and nuclei, possess intrinsic quantised angular momentum, known as spin. The spin angular momentum is described in terms of a quantum number, S, which may take integer or half integer values. The magnitude of the spin angular momentum, A_S, is given by

$$A_S = [S(S+1)]^{1/2}\,\hbar.$$

In addition, there is quantised angular momentum, A_L, associated with orbital motion of electrons in atoms. This is described in terms of a quantum number, L, which may take only integer values,

$$A_L = [L(L+1)]^{1/2}\,\hbar.$$

The total angular momentum, which is also quantised, is the resultant of these two contributions and is described in terms of a total angular momentum quantum number, J. Charged particles possessing angular momentum may be thought of as tiny circulating currents, and so have a magnetic dipole moment associated with them. The magnetic moment, $\boldsymbol{\mu}_m$, is oriented in the same direction as the angular momentum vector, $\boldsymbol{J}$, and its strength is given by

$$\mu_m = g\mu\,[J(J+1)]^{1/2},$$

where μ is the unit of magnetic moment (in the case of electrons it is the Bohr magneton, $\mu_B = 9.27410 \times 10^{-24}\ \mathrm{J\,T^{-1}}$, or for nuclei, it is the nuclear magneton $\mu_N = 5.05095 \times 10^{-27}\ \mathrm{J\,T^{-1}}$) and g is a constant, known as the spectroscopic splitting factor ($g \approx 2$ for electrons with $L = 0$, and for copper nuclei $g = 2.226$). In an applied magnetic field, the dipole experiences a torque which tends to align it in the field direction. The component of the angular momentum along the magnetic field direction is quantised and may take any one of $2J + 1$ possible values. Corresponding to each of these possible orientations of the dipole is a magnetic energy level given by

$$E = -g\mu Bj \qquad (j = -J, -J+1, \ldots, 0, \ldots, J).$$

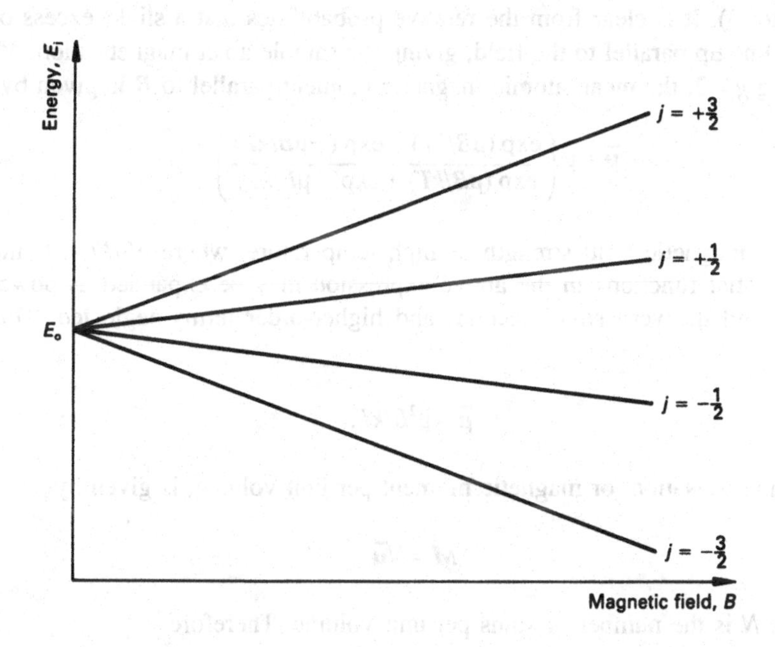

Figure 6.1 *Zeeman splitting of magnetic energy levels of an atom having total angular momentum quantum number* J = (3/2)

The difference in energy between the levels is proportional to the magnitude of the applied magnetic induction, $\boldsymbol{B}$ (Figure 6.1). The lifting of the degeneracy of the magnetic levels by an external field in this way is known as Zeeman splitting. (For a more complete introduction to atomic levels and the Zeeman effect, see Woodgate, 1970.)

Now consider a particle with spin quantum number $S = \frac{1}{2}$ and orbital quantum number $L = 0$. In the presence of an external magnetic field it may be in one of two quantum states; one for which $j = +\frac{1}{2}$, i.e. its spin points parallel to $\boldsymbol{B}$, the other has antiparallel spin, $j = -\frac{1}{2}$. The respective probabilities, $P_{\uparrow}$ and $P_{\downarrow}$, of finding the atoms in each of these states are given in terms of the statistical Boltzmann factor, $\exp(-E/kT)$:

$$P_{\uparrow} = C \exp(g\mu B/2kT),$$

and

$$P_{\downarrow} = C \exp(-g\mu B/2kT),$$

where C is a constant (this does not apply to conduction electrons in metals and semiconductors or to ^{3}He nuclei, which obey Fermi statistics; see Chapter 3). It is clear from the relative probabilities that a slight excess of spins line up parallel to the field, giving the sample a net magnetisation, $\boldsymbol{M}$. Putting $g = 2$, the mean atomic magnetic moment parallel to $\boldsymbol{B}$ is given by

$$\bar{\mu} = \mu \left(\frac{\exp(\mu B/kT) - \exp(-\mu B/kT)}{\exp(\mu B/kT) + \exp(-\mu B/kT)} \right).$$

At low magnetic field strength or high temperature, where $\mu B/kT \ll 1$, the exponential functions in the above expression may be expanded as power series and the very small second- and higher-order terms neglected. This gives

$$\bar{\mu} = \mu^2 B/kT.$$

The magnetisation, or magnetic moment per unit volume, is given by

$$\boldsymbol{M} = N\bar{\mu},$$

where N is the number of spins per unit volume. Therefore,

$$\boldsymbol{M} = N\mu^2 \boldsymbol{B}/kT.$$

Hence, the magnetisation is proportional to the strength of the applied field $\boldsymbol{B}$. A quantity known as the paramagnetic susceptibility, χ, is defined such that

$$M = \chi B/\mu_0,$$

where μ_0 is the permeability of free space;

$$\frac{\chi}{\mu_0} = \frac{N\mu^2}{kT}.$$

The $1/T$ dependence of χ is known as Curie's law. At high fields, the magnetisation reaches a saturation value corresponding to all spins lining up parallel to B:

$$M = N\mu.$$

A more general calculation of the magnetisation, along the same lines as that above, but applicable to systems having a total angular momentum quantum number other than ½, gives

$$M = Ng\mu \cdot J \cdot B_j(g\mu B/kT),$$

where $B_j(g\mu B/kT)$ is known as the Brillouin function, and is shown graphically for systems having different values of J in Figure 6.2. The

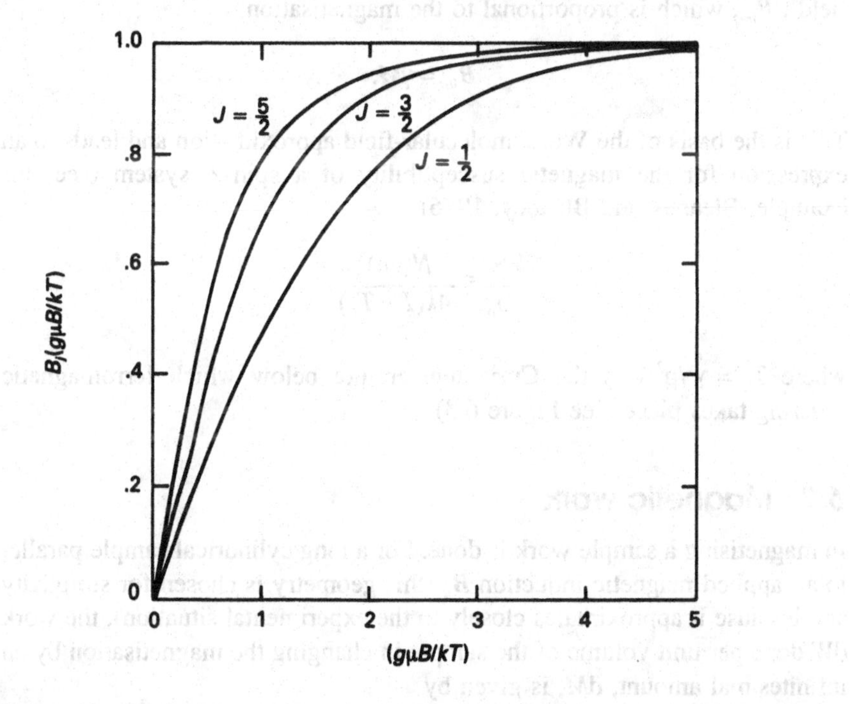

Figure 6.2 *The Brillouin function* $B_j(g\mu B/kT)$ *for different values of total angular momentum* J

corresponding low-field–high-temperature susceptibility χ is given by

$$\frac{\chi}{\mu_0} = \frac{N(g\mu)^2 \cdot J(J+1)}{3kT}.$$

It is easily confirmed that by substituting $J = \frac{1}{2}$ and $g = 2$, this becomes the same as that already derived for the spin $-\frac{1}{2}$ system. Experimental measurements of the magnetisation of various paramagnetic systems as a function B/T (see, for example, Henry, 1952) are found to be in almost perfect quantitative agreement with this theory. The Curie law behaviour at low B/T and saturation at high B/T are clearly demonstrated. It is also found that a 'high temperature' in relation to magnetic systems is really quite low by any other standards. For example, at $B = 1$ T, Curie's law is normally obeyed down to temperatures of the order 1 K.

So far, the interactions between the spins have been ignored. However, these become particularly important at low temperatures. The two most important interactions are: The dipolar interaction, which is due to the direct magnetic influence of spins on their near neighbours, and the exchange interaction, which is of quantum-mechanical origins and depends on the relative orientation of neighbouring spins. One way of taking such interactions into account is to describe their effect in terms of an 'internal field', $\boldsymbol{B}_{int}$, which is proportional to the magnetisation,

$$\boldsymbol{B}_{int} = \gamma \boldsymbol{M}.$$

This is the basis of the Weiss molecular-field approximation and leads to an expression for the magnetic susceptibility of a spin-½ system (see, for example, Bleaney and Bleaney, 1976):

$$\frac{\chi}{\mu_0} = \frac{N(g\mu)^2}{4k(T - T_C)},$$

where $T_C = \gamma N \mu^2 / k$ is the Curie temperature below which ferromagnetic ordering takes place (see Figure 6.3).

6.2 Magnetic work

In magnetising a sample work is done. For a long cylindrical sample parallel to an applied magnetic induction $\boldsymbol{B}_a$ (this geometry is chosen for simplicity and because it approximates closely to the experimental situation), the work dW done per unit volume of the sample in changing the magnetisation by an infinitesimal amount, dM, is given by

$$\mathrm{d}W = \boldsymbol{B}_a \, \mathrm{d}M.$$

A derivation of this relation will now be presented. In the chosen geometry, the magnetising field, $\boldsymbol{H}$, the magnetic induction, or flux density, $\boldsymbol{B}$, and the magnetisation, $\boldsymbol{M}$, in the sample are all parallel to $\boldsymbol{B}_a$ (Figure 6.4a), and are related by

$$\boldsymbol{B} = \mu_0(\boldsymbol{H} + \boldsymbol{M}).$$

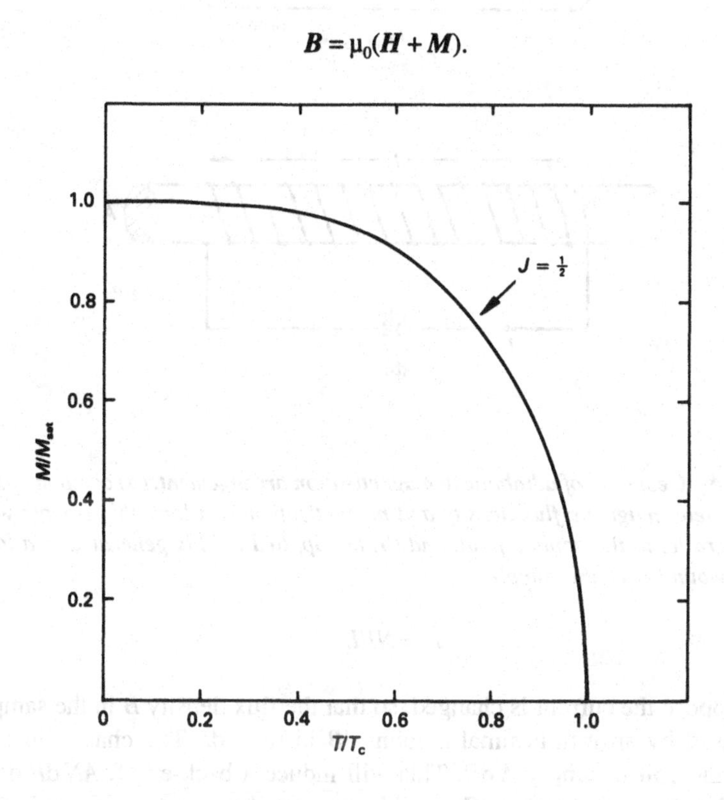

Figure 6.3 *Temperature dependence of the magnetisation of a ferromagnetic material at temperatures below the Curie temperature according to Weiss theory,* M_{sat}: *Saturation magnetisation*

Furthermore, because the tangential component of $\boldsymbol{H}$ is continuous across the sample boundary,

$$\boldsymbol{H} = \boldsymbol{H}_a = \boldsymbol{B}_a/\mu_0.$$

For the purpose of magnetising the sample, suppose that a uniform, long solenoid (by 'long' it is meant that the length l is much greater than the diameter), consisting of N turns, is tightly wound around the sample, length l and cross sectional area A (Figure 6.4b). A magnetic field, H, is established inside the coil by passing a current, I, through the windings; by Ampère's law,

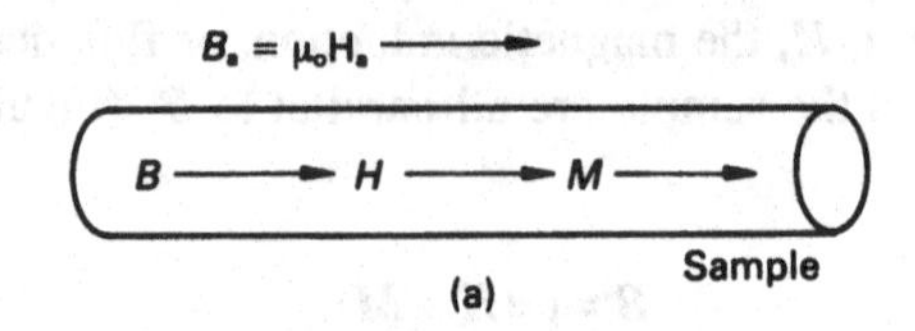

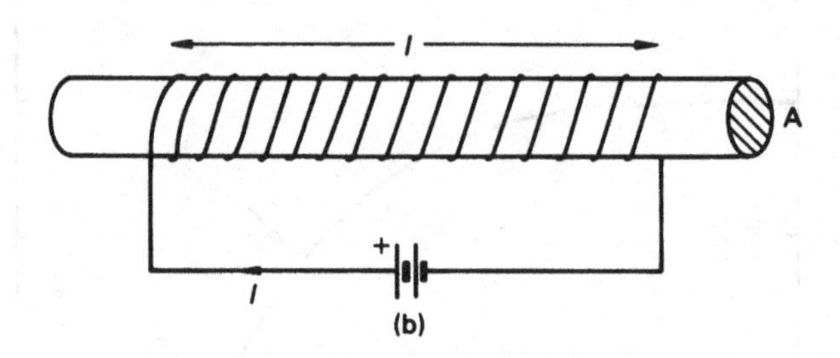

Figure 6.4 *Geometry of adiabatic demagnetisation arrangement. (a) components of magnetic field, magnetic flux density and magnetisation in a long thin sample with its axis parallel to the applied field, and (b) the applied field is generated in a long solenoid wound over the sample*

$$H = NI/l.$$

Now, suppose the current is changed, so that the flux density B in the sample is increased by an infinitesimal amount $\mathrm{d}B$ in time $\mathrm{d}t$. The change in flux through the coil is simply $A\,\mathrm{d}B$. This will induce a back-e.m.f. $AN\,\mathrm{d}B/\mathrm{d}t$ to oppose the change in current. The coil is connected to a source of e.m.f., e.g. a battery, which must do work to overcome this and maintain the current. The work done, or the power delivered from the source, is

$$AN\left(\frac{\mathrm{d}B}{\mathrm{d}t}\right)I\,\mathrm{d}t = ANI\,\mathrm{d}B.$$

Substituting Hl for NI (from the previous equation) gives

$$ANI\,\mathrm{d}B = AHl\,\mathrm{d}B,$$

and for B in terms of H and M,

$$A\cdot Hl\,\mathrm{d}B = A\cdot Hl\mu_0(\mathrm{d}H + \mathrm{d}M).$$

The work done per unit volume is obtained by dividing this by the sample volume, lA. Thus,

$$dW = \mu_0 H\,(dH + dM).$$

Bearing in mind that H is equal to H_a, the field in the absence of the sample, and that $B_a = \mu_0 H_a$, this may be expressed in the following form:

$$dW = (B_a/\mu_0)\,dB_a + B_a\,dM.$$

This is the total work done in changing the field and the magnetisation of the sample. In the absence of any sample, the work done in simply changing the field is easily shown, using the same arguments as above, to be given by the first term. Therefore, the remaining term,

$$dW = B_a\,dM,$$

is attributed to the work done in changing the magnetisation of the sample.

6.3 Entropy in magnetic sub-systems

The total entropy of a paramagnetic sample consists of a number of contributions, each of which is associated with a particular 'sub-system'. For example, a solid sample has an entropy associated with the thermal motions (phonons) in the lattice, as in the case of any non-magnetic solid. If there is a free electron gas, an entropy must be associated with this also. These may be treated quite separately from the entropy which arises from disorder in the orientation of the magnetic dipoles. The total entropy is expressed as a sum of the contributions from these various sub-systems,

$$S = S_{\text{phonons}} + S_{\text{electrons}} + S_{\text{magnetic}}.$$

In insulating solids at low enough temperatures, the lattice is essentially still, and the 'magnetic-entropy' is the dominant contribution. A measure of the magnetic ordering is the ratio of magnetic to thermal energy, $g\mu B/kT$. Consequently, the magnetic entropy is expected to depend on the ratio B/T, $S = S(B/T)$. In Chapter 1, the entropy of a system was expressed in terms of the number of distinguishable states, Ω, accessible to it. For a system of N indistinguishable particles,

$$S = Nk \ln \Omega.$$

Each particle in the magnetic sub-system could occupy any of the $2J + 1$

energy levels. The probability of occupation of the jth level is given by

$$P_j \propto \exp(g\mu Bj/kT),$$

so only at high temperatures are all $2J+1$ levels accessible. At sufficiently high temperatures the entropy of the magnetic sub-system clearly attains the limiting value S_L,

$$S_L = Nk \ln(2J+1).$$

As the temperature is reduced, the probability of occupancy of the higher-energy states is reduced, and so the entropy reduces. An alternative expression for the entropy, in terms of the statistical partition function $Z(T)$, is written

$$S = Nk\left\{\ln(Z) - \left(\frac{1}{ZkT}\right)\frac{dZ}{d(1/kT)}\right\},$$

(see for example Sprackling, 1991). The partition function is the sum over all the $2J+1$ states of their Boltzmann factors. For the case of a spin-½ system,

$$Z = \exp(-g\mu B/2kT) + \exp(g\mu B/2kT).$$

In the low-magnetic-field–high-temperature limit, i.e. $g\mu B/kT \ll 1$ (it will be shown later that this limit is applicable to magnetic cooling experiments),

$$\exp\left(\pm\frac{g\mu B}{2kT}\right) \sim 1 \pm \frac{g\mu B}{2kT}.$$

Upon substitution, this yields

$$S = Nk\left\{\ln 2 - \frac{(g\mu B)^2}{4k^2T^2}\right\}.$$

In general, for systems with spin other than ½, the entropy is given by

$$S = Nk\left\{\ln(2J+1) - \frac{g^2\mu^2 J(J+1)B^2}{3k^2T^2}\right\}.$$

This tends to the limiting value $S_L = Nk\ln(2J+1)$ as $T \to \infty$ (Figure 6.5). As the temperature is reduced towards absolute zero, the system falls into the magnetic ground state (all spins ordered up parallel to the field) and the magnetic entropy reduces to zero.

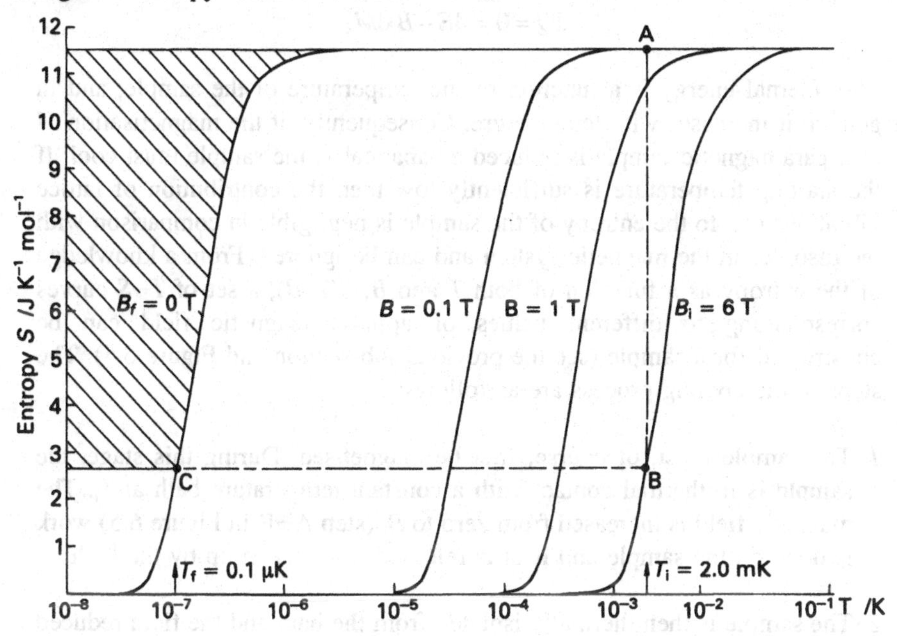

Figure 6.5 *Theoretical calculations of the nuclear magnetic entropy of copper, corresponding to different values of the applied flux density. For the equation, see the text*

6.4 Principles of magnetic cooling

It has been established that work can be done on a magnetic sample by changing its magnetisation. Just as in the case of systems on which mechanical work can be performed, it should be possible to heat or cool a sample in thermal isolation by performing magnetic work. The work done *by* the sample, when its magnetisation changes by amount dM in an external magnetic flux density B, is $-B\,\mathrm{d}M$. Application of the first law of thermodynamics to magnetic systems gives

$$dQ = T\,dS = dE + p\,dV - B\,dM.$$

Magnetic cooling is performed using solid samples for which the change in volume is negligible. Thus, negligible mechanical work is done by the sample and

$$dQ = dE - B\,dM.$$

For a thermally isolated sample,

$$dQ = 0 = dE - B\,dM.$$

The internal energy is a function of the temperature of the sample, and in general it increases with temperature. Consequently, if the magnetisation M of a paramagnetic sample is reduced adiabatically, the sample must cool. If the starting temperature is sufficiently low then the contribution of lattice vibrations etc. to the entropy of the sample is negligible in comparison with the disorder in the magnetic system and can be ignored. From a knowledge of the entropy as a function of both T and B, $S(T, B)$, a set of T–S curves corresponding to different values of applied magnetic field can be constructed for a sample (see the previous sub-section and Figure 6.5). The steps in the cooling process are as follows:

1 The sample must, of course, first be magnetised. During this stage, the sample is in thermal contact with a constant-temperature bath at T_i. The magnetic field is increased from zero to B_i (step A–B in Figure 6.5) work is done on the sample and heat is released to be taken up by the bath.

2 The sample is then thermally isolated from the bath and the field reduced from B_i to B_f. This step is taken adiabatically, i.e. $dQ = T\,dS = 0$; therefore, the entropy remains constant, B–C on the figure. Because the sample performs magnetic work as it demagnetises at the expense of its internal energy, it cools to a final temperature T_f.

In a system where there are no mutual interactions between the spins, and subject to the condition $g\mu B/kT \ll 1$ being satisfied, the entropy is simply a function of B^2/T^2. For the adiabatic (isentropic) stage,

$$S\left(\frac{B_i^2}{T_i^2}\right) = S\left(\frac{B_f^2}{T_f^2}\right),$$

where B_i, T_i and B_f, T_f are respectively the starting field and temperature and finishing field and temperature for the adiabatic stage. Therefore,

$$T_f = T_i \frac{B_f}{B_i},$$

suggesting that if $B_f = 0$, $T_f = 0$ in contradiction to the third law of thermodynamics. This does not happen in practice and the reason is that, at very low temperatures, the assumption of non-interacting spins breaks down. In the absence of an external field, the degeneracy of the magnetic levels is lifted slightly by the internal crystal fields, in a manner analogous to the Zeeman effect. This becomes important at very low temperatures, where the thermal energy kT is smaller than the energy splittings produced. The detailed nature of the interactions may be complicated. However, if they lift the degeneracy of the energy levels by ε, then it simplifies matters to define an effective internal field, b, such that it would produce a Zeeman splitting equal to ε, i.e. $b = \varepsilon/g\mu$. This is simply the magnetic field felt by a spin due to the dipolar fields of its neighbours, if there are no other interactions between the atoms or molecules in the sample. Quantitative analysis yields for the entropy under these conditions,

$$S = Nk\left\{\ln(2J+1) - g^2\mu^2 J(J+1)\left(\frac{B^2+b^2}{3k^2T^2}\right)\right\}.$$

Therefore, it follows that

$$T_f = T_i\left(\frac{B_f^2+b^2}{B_i^2+b^2}\right)^{1/2},$$

and in the limit $b \ll B_i$ this reduces to

$$T_f = T_i\, b/B_i,$$

at $B_f = 0$ T.

EXAMPLE

It is intended to cool a copper sample by adiabatic nuclear demagnetisation from a starting field of 3 T and starting temperature of 10 mK. The internal dipolar field is 0.3 mT; what is the final temperature achieved?

Using the equation

$$T_f = T_i\, b/B_i = 0.3 \times 10^{-3} \times 10 \times 10^{-3}/3 = 10^{-6}\,\mathrm{K},$$

that is, the final temperature is 1.0 μK, a very low temperature by any standards! It is interesting to note that in spite of this, the condition $g\mu b/kT < 1$ is still satisfied. For copper $g = 2.2$, therefore, at $b = 0.3$ mT and $T = 1$ μK,

$$g\mu b/kT = 0.25,$$

which is less than unity.

It is one thing to be able to cool nuclear spins to such low temperatures by adiabatic demagnetisation, but for the technique to be of any practical use it is also necessary to cool the lattice to which the experiment can be attached. The lattice temperature of the paramagnetic sample will not necessarily follow closely the spin temperature during the demagnetisation step. It takes time for the exchange of energy between the spins and the lattice, and hence to establish thermal equilibrium between them. The mechanism of energy exchange is the so-called *spin—lattice coupling*; it arises owing to the effect of thermal motions in the lattice on the magnetic environment of the spins. In metals, the spin—lattice coupling is mediated by the conduction electrons, which makes the process more efficient than in insulating materials. The timescale for this process to establish thermal equilibrium is called the *spin—lattice relaxation time*, τ_1, which is given by the Korringa relation,

$$\tau_1 = C_K/T_e,$$

where T_e is the electron temperature. The Korringa constant, C_K, is weakly field dependent, having a value of 1.1 K s in copper at fields above 0.3 mT. In copper, the spin—lattice relaxation time is about 1000 s at $T_e = 1$ mK. In non-metals, the spin—lattice relaxation time can be hours or even days at comparable temperatures. This is why metals are used in nuclear cooling systems.

In a constant external field the rate of change of nuclear spin temperature is proportional to the difference between the nuclear spin and the electron temperatures, T_n and T_e, respectively, and is given by

$$\frac{\mathrm{d}}{\mathrm{d}t}\left(\frac{1}{T_n}\right) = -\frac{1}{\tau_1}\left(\frac{1}{T_n} - \frac{1}{T_e}\right).$$

It is reasonable to consider the nuclear and electron systems as having separate temperatures in this way, because the time taken for each system to reach internal equilibrium is generally less than the time taken for them to come into equilibrium with each other. Energy is conserved in the sample if it is thermally isolated from the surroundings, hence the rate of change of T_e can be related to the rate of change of T_n by

$$c_e \frac{dT_e}{dt} + c_n \frac{dT_n}{dt} = 0,$$

where c_e and c_n are the specific heat capacities of the electronic and nuclear systems, respectively. Combining this with the expression for the rate of change of T_n, above, gives, after a little rearrangement,

$$\frac{dT_e}{dt} = \frac{T_n c_n}{c_e C_K}(T_n - T_e).$$

Now it is possible to estimate the demagnetisation rate necessary to maintain a constant small temperature difference, ΔT, between the nuclear spins and electrons. For a constant temperature difference

$$\frac{dT_e}{dt} = \frac{dT_n}{dt} = \left(\frac{T_i}{B_i}\right)\frac{dB}{dt},$$

where $T_n/B \approx T_i/B_i$ has been used and the small internal field ignored. The specific heat capacity of conduction electrons in copper is $\gamma T_e = 7.0 \times 10^{-4} T_e$ J K^{-1} mol^{-1} (see Section 2.2.2). The specific heat of the nuclear spin system may be obtained from its entropy, and for $J = 3/2$,

$$c_n = T_n\left(\frac{dS}{dT}\right)_B = \frac{5N(g\mu B)^2}{2kT_n^2}.$$

If ΔT is small, $T_e \approx T_n$, and

$$\frac{dB}{dt} = \frac{5Ng^2\mu^2\Delta T}{2k\gamma C_K}\left(\frac{B_i}{T_i}\right)^3.$$

Therefore, to maintain a temperature difference of just 1 nK in the experimental situation described in the example above, a demagnetisation rate of 0.00047 T s^{-1} would be needed, giving a total demagnetisation time of 106 min. If the time taken to demagnetise the copper sample is comparable to, or longer than this, and it normally is, then it is reasonable to assume that the electron and hence lattice temperature follows very closely the nuclear spin temperature.

It is quite straightforward to estimate the final temperature of the copper sample under the assumption that the spins and lattice are in equilibrium during demagnetisation. The entropy of the sample is given by

$$S = Nk\left\{\ln(4) - \frac{5g^2\mu^2}{4}\left(\frac{B^2+b^2}{k^2T^2}\right)\right\} + \gamma T,$$

where the last term on the right is the contribution from the electrons. The lattice term is minuscule at such low temperatures and has been neglected. The demagnetisation step is taken isentropically, $\Delta S = S_f - S_i = 0$, and hence under the same conditions as in the example the final temperature turns out to be a mere 10^{-5} μK above the minimum temperature that would be obtained if the spins were cooled in complete isolation.

An alternative scheme would be to demagnetise the spins rapidly so that the electron temperature could not keep up with the spin temperature. This might be the case where very low spin temperatures are sought and the corresponding values of τ_1 become very long. Immediately after demagnetisation the spins are at T_f and the electrons are left at T_i. Over a time of the order of τ_1 the electrons will exchange heat with the nuclei, the lattice will cool and the nuclear spin temperature will increase until they reach equilibrium at T_{eq}. Again, energy is conserved in the thermally isolated sample, hence

$$\int_{T_i}^{T_{eq}} c_e \, dT_e + \int_{T_f}^{T_{eq}} c_n \, dT_n = 0.$$

Evaluating the integrals in the case of copper gives, for $T_{eq} \ll T_i$,

$$\frac{1}{T_{eq}} = \frac{1}{T_f} - 26\frac{T_i^2}{B_f^2}.$$

Which means that if $T_i = 10$ mK, $T_f = 1$ μK and $B_f = 0.3$ mT, then the whole sample would reach equilibrium at a temperature of 1.03 μK. Even under such extreme conditions the increase in temperature is hardly significant owing to the relatively small heat capacity of the electrons.

Of course, all of the above depends on the assumption that any heat leaks are negligible. If not, the situation becomes more complicated (see, for example, Betts, 1989).

As mentioned earlier, adiabatic demagnetisation cooling is a one-shot process, in which the temperature is reduced to the base temperature T_f, after which the system warms up slowly. The time taken to warm depends on the balance between the heat leaks into the system and the thermal capacity of the nuclear spin system. In the process of warming up from T_f to T_i the nuclear spin system absorbs heat,

$$\Delta Q = \int_{S_f}^{S_i} T \, dS,$$

where S_f and S_i are the entropies of the nuclear spins at T_f and T_i, respectively. This is the area of the shaded portion in Figure 6.5. The amount of heat absorbed is strongly dependent on the final value of the field. In the example shown in Figure 6.5, the externally applied field is reduced to zero to obtain the lowest final temperature; upon warming the spins can absorb 2.4 μJ per mole of copper. However, if the field was reduced to 100 mT, the base temperature would (only!) reach 32 μK, but the thermal capacity would increase to about 0.8 mJ per mole.

6.5 Practical magnetic cooling

Regardless of whether it is by the demagnetisation of electronic or of nuclear spins that cooling is to be brought about, the ideal paramagnetic sample should have the following properties:

1 The internal field should be small.

2 The heat capacity associated with the magnetic system should be large.

3 There should be an effective mechanism of spin—lattice coupling and a short spin—lattice relaxation time.

4 The lattice specific heat should be small.

The internal field can be reduced by diluting the magnetic atoms, but this reduces the magnetic heat capacity; a compromise must be sought between these two requirements. Cerium magnesium nitrate (CMN) is often used for cooling by adiabatic demagnetisation of electronic spins. It has an internal field of about 5 mT, about an order of magnitude less than the alternatives. The sample is mounted in a vacuum chamber whose walls are maintained at near 1 K by a pumped ^{4}He pot (Figure 6.6). Low-pressure, ^{4}He, exchange gas is let into the chamber to maintain the sample at 1 K, while the field produced by a superconducting solenoid surrounding the sample space is ramped up. The exchange gas is then pumped away to thermally isolate the sample and the field swept down. If the maximum field was 2 T then a final temperature of 2.5 mK may be reached.

The nuclear magnetic moment and, therefore, the internal dipolar field are some 2000 times weaker than the electronic moment, allowing much lower temperatures to be reached from a starting temperature in the mK range. Copper is a popular choice: it has a number of advantages over its rivals, which include the following: (1) It is readily available in good quantities and is easily fabricated into the required form; (2) the spin–lattice coupling via the conduction electrons is good; (3) the internal fields are very small (<0.3 mT), and (4) no ferromagnetic or antiferromagnetic ordering takes place at temperatures above 100 nK. An alternative choice of sample for nuclear cooling is the intermetallic compound $PrNi_5$ (praseodymium nickel five). This material is an example of a hyperfine-enhanced paramagnet (Al'tshuler, 1966), in which the application of an external field modifies the electronic configuration to produce a greatly enhanced field at the nucleus (some 14 times stronger than the applied field in this case). Because it magnetically orders at about 0.4 mK, nuclear refrigerators based on $PrNi_5$ cannot achieve the lowest base temperatures available using copper. However, temperatures around 1 mK can be obtained using smaller and therefore cheaper magnets.

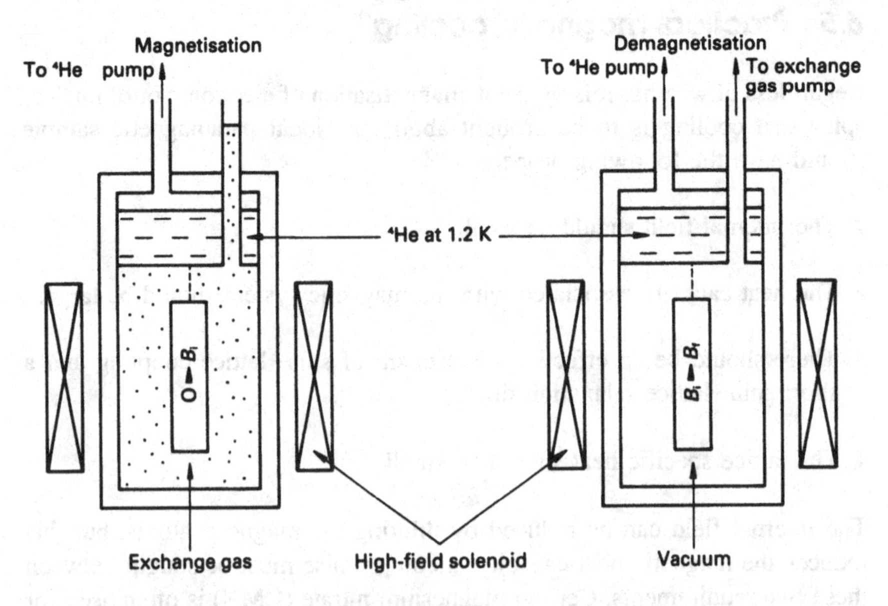

Figure 6.6 *The stages in cooling a sample by adiabatic demagnetisation*

Figure 6.7 shows a single-stage nuclear cooling cryostat using copper. To minimise eddy current heating when the field is ramped, the copper sample is made from a bundle of many fine wires. They are positioned in the bore

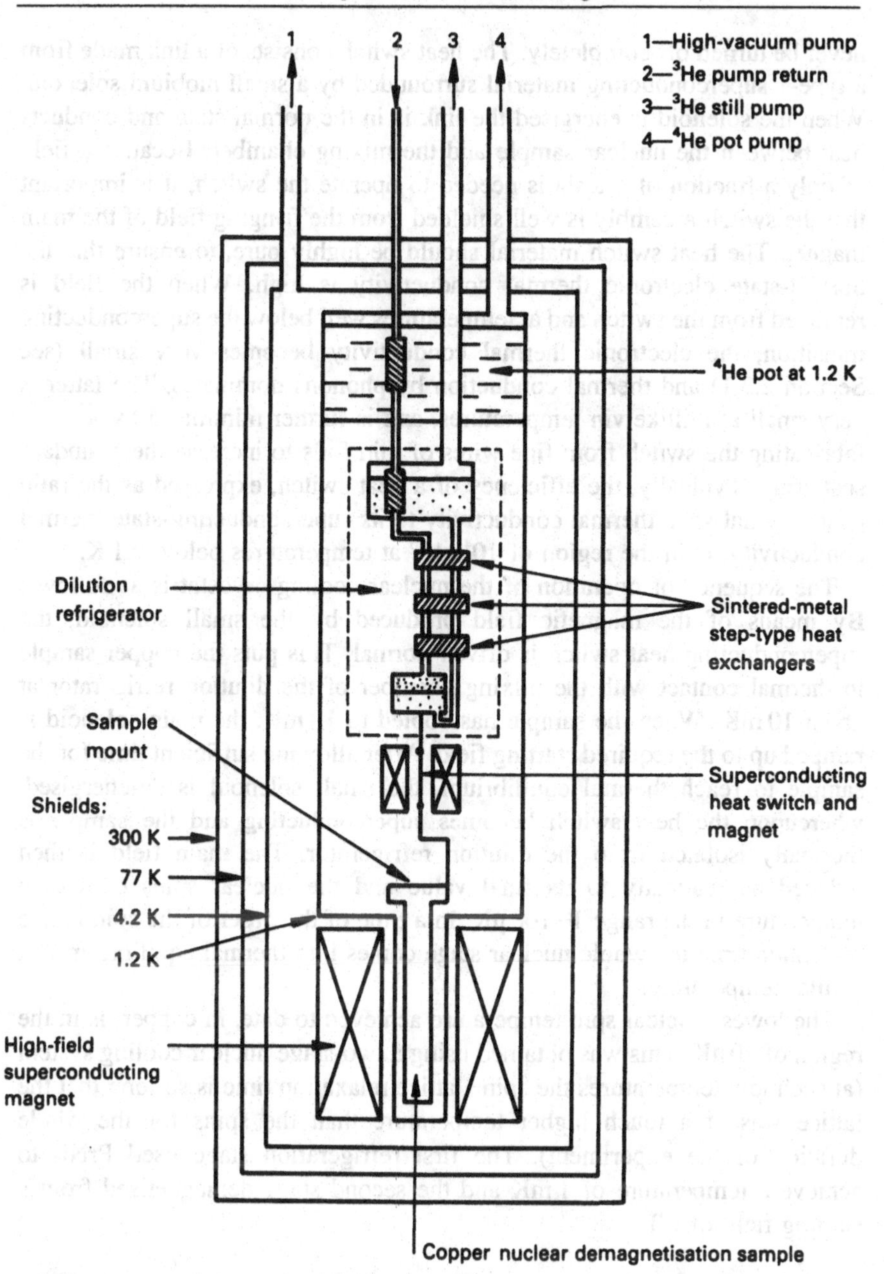

Figure 6.7 *An example of a single-stage nuclear cooling system used to obtain sub-microkelvin temperatures*

of a high field (>5 T) superconducting solenoid and in thermal contact with the mixing chamber of a dilution refrigerator via a superconducting heat switch. Superconducting heat switches have the advantage that they generate very little heat in operation. However, the main disadvantage is that they can never be turned off completely. The heat switch consists of a link made from a type-1 superconducting material surrounded by a small niobium solenoid. When the solenoid is energised the link is in the normal state and conducts heat between the nuclear sample and the mixing chamber. Because a field of only a fraction of a tesla is needed to operate the switch, it is important that the switch assembly is well shielded from the fringing field of the main magnet. The heat switch material should be highly pure, to ensure that the normal-state electronic thermal conductivity is high. When the field is removed from the switch and at temperatures well below the superconducting transition, the electronic thermal conductivity becomes very small (see Section 2.3.1) and thermal conduction by phonons dominates. The latter is very small at millikelvin temperatures, and is further minimised by fabricating the switch from fine wires of thin foils to increase the boundary scattering. Typically, the efficiency of a heat switch, expressed as the ratio of its normal-state thermal conductivity to its superconducting-state thermal conductivity, is in the region of 10^4–10^6 at temperatures below 0.1 K.

The sequence of operation of the nuclear cooling cryostat is as follows: By means of the magnetic field produced by the small solenoid, the superconducting heat switch is driven normal. This puts the copper sample in thermal contact with the mixing chamber of the dilution refrigerator at about 10 mK. When the sample has cooled to 10 mK, the main solenoid is ramped up to the required starting field. After allowing sufficient time for the sample to reach thermal equilibrium, the small solenoid is de-energised, whereupon the heat switch becomes superconducting and the sample is thermally isolated from the dilution refrigerator. The main field is then reduced adiabatically to its final value and the nuclear spins cool to a temperature in the range 1–100 μK. In a time of the order of the spin lattice relaxation time the whole nuclear stage comes into thermal equilibrium at a similar temperature.

The lowest nuclear spin temperature achieved to date, in copper, is in the region of 50 nK. This was obtained using a two-stage nuclear cooling system (at such low temperatures the spin–lattice relaxation time is so long that the lattice was at a much higher temperature than the spins for the whole duration of the experiment). The first refrigeration stage used $PrNi_5$ to achieve a temperature of 1 mK and the second stage demagnetised from a starting field of 7 T.

6.6 Working at microkelvin temperatures

It is very difficult indeed to perform experiments in the microkelvin regime, and only a few laboratories ppssess the necessary expertise. Although capable of reaching extremely low temperatures, nuclear magnetic refrigerators are often used in the temperature range 0.1 to 1 mK. A typical application would be to cool a cell containing ^{3}He so that experiments on the superfluid phases of ^{3}He could be performed. To maximise the low-temperature hold time, the measures outlined in Section 5.4 to restrict the stray heat input to the refrigerator must be applied even more rigidly. In addition, there is a serious risk of heating caused by the eddy currents induced in metal parts of the refrigeration system during the demagnetisation step. This needs to be taken account of in the design of the whole cryostat and not just in that of the copper sample. The main difficulty with nuclear refrigeration is the rather tortuous heat transfer route between the experimental sample and the spins. As has already been mentioned, in the case of copper the lattice is in contact with the spins via the conduction electrons and the lattice cools towards the spin temperature in times of the order of the spin–lattice relaxation time. The experimental sample is in thermal contact with the copper lattice, and so there will also be a boundary resistance to consider. For example, if the sample is a cell of ^{3}He, there is a large Kapitza resistance. This is partially overcome by using a sintered metal cell to increase the thermal contact area to the liquid helium. Any measurements on the sample, such as thermometry, must be performed at nanowatt power levels, to avoid raising the sample temperature well above the spin temperature. Details of thermometry suitable for use at such low temperatures are given in the next chapter. With good design, copper nuclear refrigerators can hold the sample temperature below about 2 μK for over 100 hours.

When matter is cooled to sub-microkelvin temperatures, many physical processes freeze out completely. The lattice has long since become stationary (apart from any zero point motions), electrons are in their ground state and many metals have become either superconducting or magnetically ordered. In some materials it is only the nuclear spins whose properties are still dependent on temperature. So why does the quest for still lower temperatures carry on in some laboratories? It could be answered that as long as it is possible to reduce the temperature of something, there must be a physical property that is still showing some temperature dependence and, as such, is worth studying.

Bibliography

Al'tshuler, S. A. (1966). *Soviet Physics–JETP*, **3**, 112

Betts, D. S. (1989). *An Introduction to Millikelvin Technology* (Cambridge: Cambridge University Press)

Bleaney, B. I. and Bleaney, B. (1976). *Electricity and Magnetism* (Oxford: Oxford University Press)

Debye, P. (1926). *Ann. Phys.*, **81**, 1154
Giauque, W. F. and MacDougall, D. P. (1933). *Phys. Rev.*, **43**, 768
Henry, W. E. (1952). *Phys. Rev.*, **88**, 561
Sprackling, M. (1991). *Thermal Physics* (London and Basingstoke: Macmillan)
Woodgate, G. K. (1970). *Elementary Atomic Structure* (London: McGraw–Hill)

7

Thermometry

In experiments on matter at low temperatures, some means of accurate temperature measurement is necessary to quantify the results fully so that they can be compared with the appropriate theoretical models. Thermometers have some property, preferably one that is easily measured, which changes with temperature in a known or predictable way. The basic requirements for thermometers, the need for good thermal contact to the apparatus or sample under investigation, low self-heating and fast response to changes in temperature etc. are essentially the same at low temperatures as at higher temperatures. However, some of the technical difficulties in fulfilling these requirements increase as temperatures are reduced. Quite a large number of different types of thermometer are used at low temperatures, depending on temperature range, the required accuracy, size and cost. However, all thermometers fall into one of two categories: (1) primary or absolute thermometers or (2) secondary thermometers.

Primary thermometers are those for which some measurable physical property is related to temperature in a way that can be predicted theoretically, in terms of either fundamental physical constants or other temperature-independent quantities that can be determined by experimental measurement. Primary thermometers should not normally require calibration. However, in practice, calibration at one or two fixed points is sometimes necessary to account for small corrections to the ideal theoretical characteristics.

Secondary thermometers possess a measurable property which varies with temperature in a reproducible way over the required range. They must be calibrated using primary thermometers, but have the advantage that they are often more convenient and easier to use than primary thermometers. Some secondary thermometers are so reliable that they are supplied calibrated and all they require is a single point check to ensure accurate measurements.

Others need to be calibrated every time they are used.

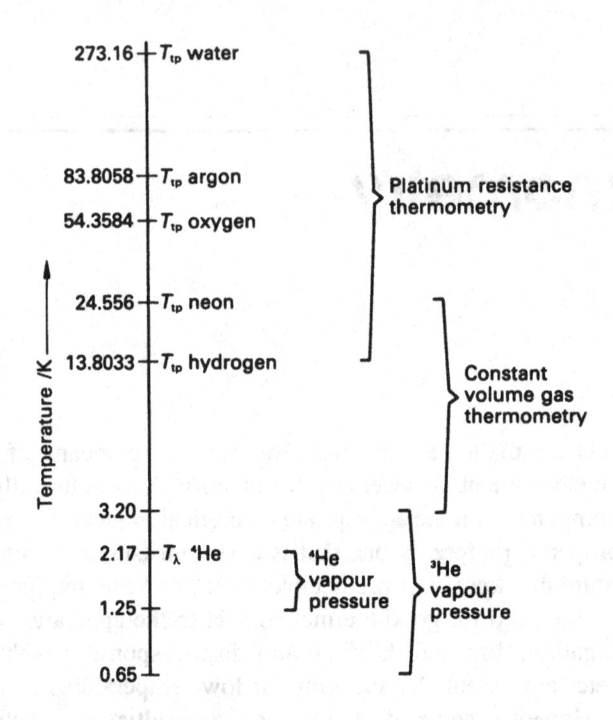

Figure 7.1 *The low-temperature part of ITS-90 (0.65 K–273.16 K), showing some of the fixed points that define the scale and the interpolation methods. Detailed instructions for the construction and calibration of the interpolation instruments are given in the specifications of the scale.* T_{tp}: *triple point temperature*

To facilitate direct comparison of experimental results obtained in different laboratories, it is desirable to relate thermometry measurements to internationally agreed temperature scales. Such scales should reproduce the true thermodynamic (Kelvin) scale, which is defined in terms of the perfect gas thermometer. This is the work of standards laboratories around the world. At the time of writing, the International Temperature Scale of 1990 (ITS-90) is in force (Quinn, 1990). It extends from 0.65 K to over 2000 K. ITS-90 replaced the International Practical Temperature Scale of 1968 (IPTS-68), which only extended down to 13.81 K and the Provisional Temperature Scale of 1976 (EPT-76), which covered 0.5 to 30 K. ITS-90 is based on a number of fixed points and is extended to lower temperatures using constant-volume gas thermometers and helium vapour pressure thermometry (see Sections 7.1.1 and 7.1.2 and Figure 7.1). Scales like ITS-90 can be made transportable by calibrating stable secondary thermometers against the primary thermometers used to define the scale in the standards laboratories. Currently, there is no consensus on the basis of an extension of ITS-90 into the millikelvin range, although there are a number of reliable primary techniques for use in this range (see Section 7.2).

7.1 Thermometry between 1 K and room temperature

The choice of thermometer for any particular temperature range is based on the following considerations: The property which is used to indicate temperature must be measurable with good accuracy and with relative ease. It should be sensitive, i.e. it should exhibit relatively large percentage changes with variations in temperature over the required measurement range. Figure 7.2 shows the different types of thermometer used between 1 K and room temperature. The list is not exhaustive but contains the more popular types, which will now be considered in more detail.

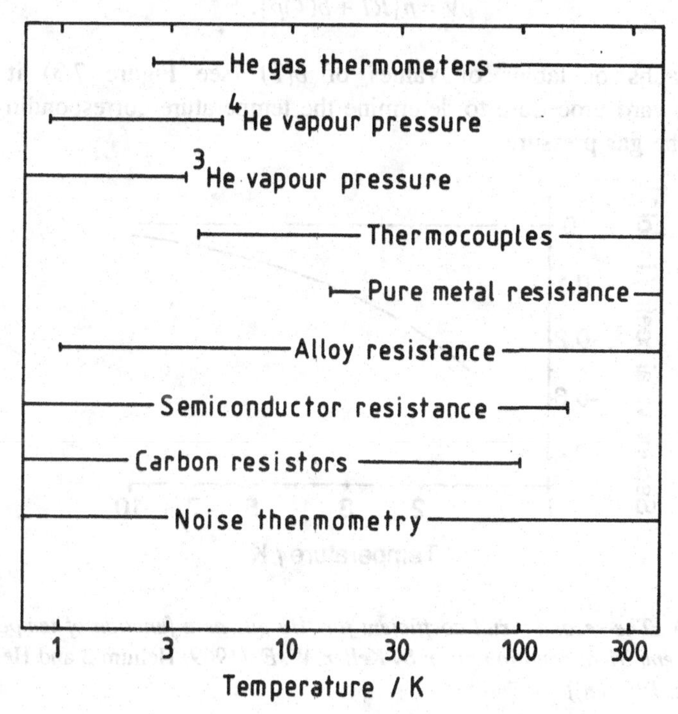

Figure 7.2 *Thermometers suitable for use between room temperature and 1 K; the approximate range over which each type may be used is indicated*

7.1.1 Gas thermometers

An ideal gas would form the basis of a perfect primary thermometer, as discussed in Chapter 1. Over the whole range of temperature right down to

0 K, its pressure at constant volume is linearly proportional to the absolute temperature. However, the ideal gas exists in theory only; in the case of a real gas the pressure–temperature relationship deviates from this ideal law. Helium gas approximates most closely to ideal behaviour at temperatures above about 10 K and is, therefore, a good choice for use in gas thermometers. At lower temperatures, where the interatomic forces become important, the following equation of state applies:

$$pV = n\{RT + b(T)p + c(T)p^2 + d(T)p^3 + \cdots\},$$

where b, c and d are the temperature-dependent virial coefficients. In the case of helium, c, d and the higher-order coefficients are very small and may be neglected, leaving

$$pV = n\{RT + b(T)p\}.$$

Using graphs or tables of values of $b(T)$ (see Figure 7.3) it is a straightforward procedure to determine the temperature corresponding to a value of the gas pressure.

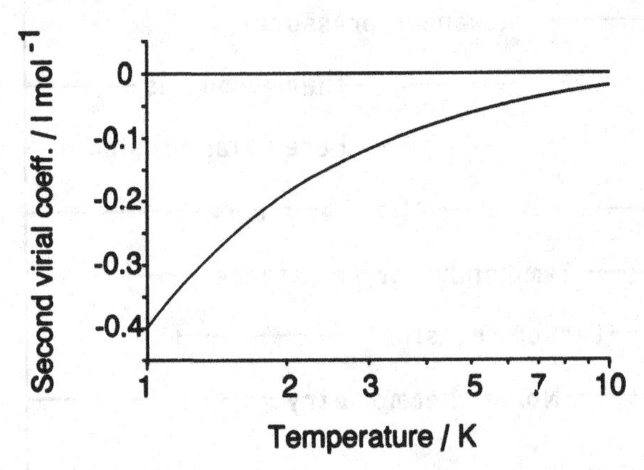

Figure 7.3 *The second virial coefficient for ^{4}He gas as a function of temperature [Based on empirical equation given by Keller, W. E. (1969)* Helium-3 and Helium-4 *(New York: Plenum)]*

A practical gas thermometer consists of a bulb of helium gas connected via a capillary tube to a constant-volume pressure-measuring device (see Figure 7.4). A system of pipes and valves allows the thermometer to be evacuated using a vacuum pump before a charge of clean gas is admitted. For the greatest accuracy, the bulb would ideally have a volume very much greater than the total volume of the pressure-measuring device and the tube added together. Often this is not possible, because space in cryogenic apparatus is limited and so correction factors need to be included. Good thermal contact is ensured by using a bulb made of copper, or another metal of high thermal conductivity, which is firmly clamped or soldered to the apparatus. To obtain measurements with an accuracy of greater than a few millikelvin at 4 K, additional correction factors are used. These take account of such effects as thermal expansion and contraction of the gas bulb and the variation in temperature of the pressure-measuring device.

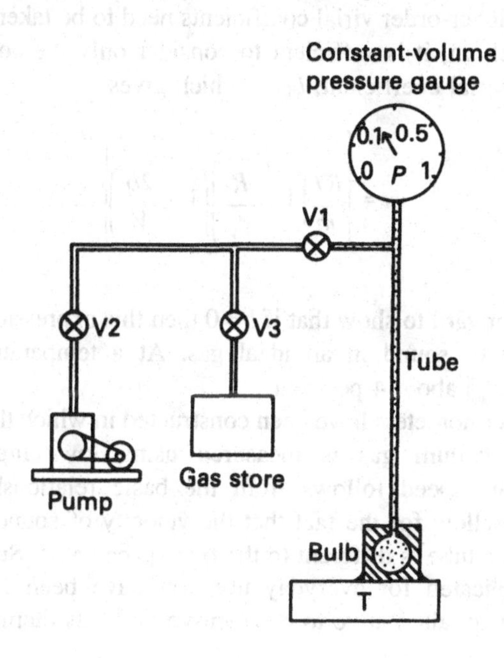

Figure 7.4 *A constant-volume gas thermometry system*

If all of the relevant correction factors are taken into account, a well-designed gas thermometer can operate as a satisfactory primary standard. However, it is usually easier to calibrate such a gas thermometer at two fixed temperatures which could, for example, be the boiling points of nitrogen and of ^{4}He at atmospheric pressure. Following such calibration, all other temperatures can be calculated.

The gas thermometer is a rather cumbersome instrument and, owing to the length of capillary between the bulb and the manometer, has a rather slow response. Furthermore, the minimum temperature that can be measured by one is about 2.5 K. For these reasons, gas thermometers are not widely used in low-temperature laboratories, although optimised ones have been used to establish international temperature scales against which other thermometers are calibrated.

The speed of sound is another temperature-dependent property of gases that can be used in thermometry. The speed of sound in an ideal gas at moderate pressure is given by

$$u = (\gamma RT/m)^{1/2},$$

where R is the gas constant, m the molecular weight of the gas and γ the ratio of the specific heat at constant pressure to that at constant volume. In a real gas at high pressures or low temperatures, contributions from the second- and higher-order virial coefficients need to be taken into account. In the case of helium, it is sufficient to consider only the contribution of the second-order virial coefficient, $b(T)$, which gives

$$u = \left\{\frac{RT}{m}\left(1 + \frac{R}{c_V}\right)\left(1 + \frac{2b}{V}\right)\right\}^{1/2}.$$

It is straight forward to show that if $b=0$ then this expression reduces to that for the speed of sound in an ideal gas. At a temperature of 10 K the correction to u is about 4 per cent.

Practical thermometers have been constructed in which the wavelength, λ, of sound in helium gas is measured using low-temperature acoustic resonators. The speed follows from the basic relationship $u = f\lambda$, with corrections to allow for the fact that the velocity of sound waves confined by walls or in a tube is different to the free-space value. Such thermometers are too complicated for everyday use, but have been used in standards laboratories as an alternative to constant-volume gas thermometers.

7.1.2 Vapour pressure thermometry

Vapour pressure thermometry uses the well-known property that the equilibrium vapour pressure of a substance increases with temperature. It is particularly useful over the ranges of temperature that can be achieved by pumping on liquid helium (see Chapter 4 and Table 7.1) and is often used in conjunction with that method of cooling. The relationship between vapour pressure and temperature of a substance is given by the Clausius–Clapeyron equation,

$$\frac{dp}{dT} = \frac{L(T)}{T\,\Delta V},$$

where $L(T)$ is the latent heat of vaporisation of the substance and ΔV the change in volume. The change in volume when the substance evaporates at temperature T is simply taken to be the volume of the gas liberated under the

Table 7.1 ^{3}He and ^{4}He vapour pressures at a few selected temperatures

T/K	Vapour pressure/Pa ^{3}He	^{4}He
4.201	–	100 000
4.000	–	82 220
3.500	–	47 440
3.177	100 000	31 310
3.000	82 660	24 280
2.500	44 440	10 330
2.174	27 240 (λ-point)	5 070
2.000	20 180	3 170
1.750	12 320	1 380
1.500	6 760	480
1.250	3 190	117
1.000	1 160	17
0.750	261	–
0.500	20	–

prevailing conditions of temperature and pressure (the reduction in volume of the liquid or solid is relatively small and can be ignored). Assuming the gas to be ideal, $\Delta V = RT/p$ per mole evaporated, so that integrating gives

$$R \ln p + const. = \int \frac{L(T)}{T^2} \mathrm{d}T.$$

With a knowledge of $L(T)$, this expression can be evaluated to give an absolute scale for the temperature in terms of vapour pressure. In practice, a set of secondary vapour pressure scales are used: For example, the T_{58} ^{4}He scale (58 stands for the year 1958, when the scale was adopted) and the T_{62} ^{3}He scale. In ITS-90, these scales have been superseded by the relation,

$$T = \sum_i A_i \left\{ \frac{(\ln p + B)}{C} \right\}^i$$

(Rusby and Durieux, 1984), where the values of the constants A_i, B and C

are listed in Table 7.2 and the vapour pressure, p, is in Pa. The maximum discrepancy between ITS-90 and the older scales is only 8 mK at 5 K.

Table 7.2 Values of the coefficients in the equation defining the ITS-90 helium vapour pressure thermometry scale [after Quinn, 1990]

Coefficient	^{4}He 1.25–2.1768 K	^{4}He 2.1768–5.00 K	^{3}He 0.65–3.20 K
A_0	1.392408	3.146631	1.053447
A_1	0.527153	1.357655	0.980106
A_2	0.166756	0.413923	0.676380
A_3	0.050988	0.091159	0.372692
A_4	0.026514	0.016349	0.151656
A_5	0.001975	0.001826	-0.002263
A_6	-0.017976	-0.004325	0.006596
A_7	0.005409	-0.004973	0.088966
A_8	0.013259	0	-0.004770
A_9	0	0	-0.054943
B	5.6	10.3	7.3
C	2.9	1.9	4.3

Figure 7.5 shows a practical vapour pressure thermometry system. The pressure-measuring device, i.e. a mercury or oil manometer (there is no need to use a constant-volume device, as in gas thermometry), is connected via a narrow tube to the pumped helium pot or to a small bulb containing liquid helium that is in close thermal contact with the apparatus. Care must be taken when using ^{4}He at temperatures below the lambda point to avoid superfluid film flow along the connecting tube, which is responsible for heat leaks to the helium bath. The lower limit to the temperature that can be measured by vapour pressure thermometry is set by the lowest pressure that can be measured accurately. Using a light oil (butyl phthalate) manometer, this is about 10 Pa, corresponding to 1 mm of oil, which gives lower temperature limits of 0.9 K and 0.4 K in the case of ^{4}He and ^{3}He, respectively. Corrections that must be taken into account include the following:

1 The thermal boundary (Kapitza) resistance between metal apparatus and the liquid helium (see Section 4.7). This will result in the apparatus being at a slightly higher temperature than the helium, if the experiment generates any appreciable amount of heat.

2 The thermomolecular pressure difference which exists between two connected volumes of gas at different temperatures, where the connecting tube has a diameter less than or comparable to the mean free path of the gas molecules. This problem is accentuated at lower pressures, where the mean free path is long. It is generally impractical to use a tube of sufficiently large diameter to make the effect negligible at the lowest temperatures in the range, and therefore tables of the thermomolecular pressure correction have been compiled (Roberts and Sydoriak, 1957).

3 A higher vapour pressure is required to form bubbles below the surface of a boiling liquid, owing to the hydrostatic pressure at depth. The liquid boiling below the surface is therefore at a slightly higher temperature than that at the surface where the vapour pressure is measured.

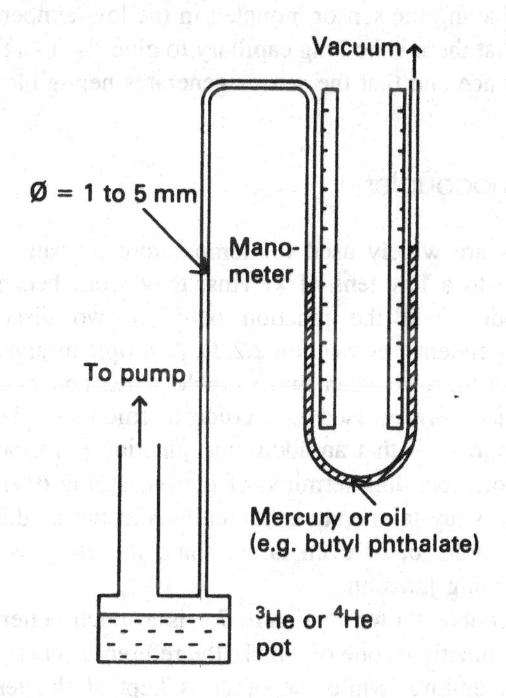

Figure 7.5 *Helium vapour pressure thermometry system*

It should be appreciated that accurate vapour pressure thermometry is time-consuming. It is important to allow sufficient time for the liquid and vapour to reach equilibrium at the same temperature. This does not take long if measurements are made while pumping the helium pot to reduce its temperature. More time is required if the apparatus is warming up when the measurement is made: When the pumping line is closed off to allow the system to warm up, it is found that the pressure in the pot will rise immediately as the liquid evaporates. However, because of its large heat capacity the liquid helium may still be at a much lower temperature than is indicated by the vapour pressure. It can take a long time for the small heat leaks into the pot to warm the liquid. In many laboratories, vapour pressure thermometry is used as the standard against which other more convenient forms of thermometry are calibrated during the first cool-down of the apparatus.

A development of vapour pressure thermometry capable of achieving a temperature resolution of the order of one nanokelvin uses a capacitive pressure transducer which is mounted in the low-temperature part of the apparatus (Steinburg and Ahlers, 1983). The gas pressure acts on a thin diaphragm that forms both one wall of the liquid-containing cell and one plate of a parallel-plate capacitor. Capacitance can be measured very accurately using an A.C. capacitance bridge external to the cryostat. The advantages of having the sensor mounted in the low-temperature part of the apparatus are that there is no long capillary to give rise to a thermomolecular pressure difference and that the sensor generates negligible thermal noise.

7.1.3 Thermocouples

Thermocouples are widely used for temperature measurement from many hundreds down to a few tens of kelvins; they work because the potential difference produced at the junction between two dissimilar metals is temperature-dependent (see Section 2.2.3). A simple arrangement consisting of a sensing junction between two suitable wires connected to a voltage-measuring device may be used as a crude thermometer. The problem with such an arrangement is that an additional junction is formed between each wire and the corresponding terminal of the measuring device. The changes in e.m.f. due to stray temperature fluctuations at these additional junctions may be comparable to, or even larger than, the changes in e.m.f. at the temperature-sensing junction.

The thermocouple shown in Figure 7.6 is a much better arrangement; it consists of two junctions, one of which, the reference junction, is maintained at a fixed temperature, while the other is kept at the temperature to be measured. The measuring device is inserted into a break in one of the wires, and measures the Seebeck e.m.f. corresponding to the temperature difference of the two junctions. Since the wires connected to each terminal of the voltmeter are of the same type, the stray thermal e.m.f.s will balance out, provided both terminals are at the same temperature. It is advantageous to maintain the reference junction at a temperature close to that being measured. For example, if a thermocouple is to be used to measure temperatures around

4.2 K, the reference junction could be maintained at 4.2 K by immersion in liquid ^{4}He boiling at atmospheric pressure. This avoids the problem of having to measure small changes in a large standing e.m.f. The thermocouple e.m.f. approximates in its temperature dependence to the power series

$$E = a_0 + a_1\,\Delta T + a_2\,\Delta T^2 + \cdots,$$

where ΔT is the temperature difference between the two junctions. If ΔT is small, then the first three terms are a sufficiently good approximation.

The practice of using calibration equations of the above type to describe the characteristics of thermocouples and also other types of temperature sensor, e.g. the resistance thermometers described in the next section, is widespread. Such equations allow for more accurate interpolation between calibration points. Often, different thermometers of the same family follow the same basic equation except for different numerical coefficients. This behaviour may be exploited to minimise the number of points required to calibrate a particular device.

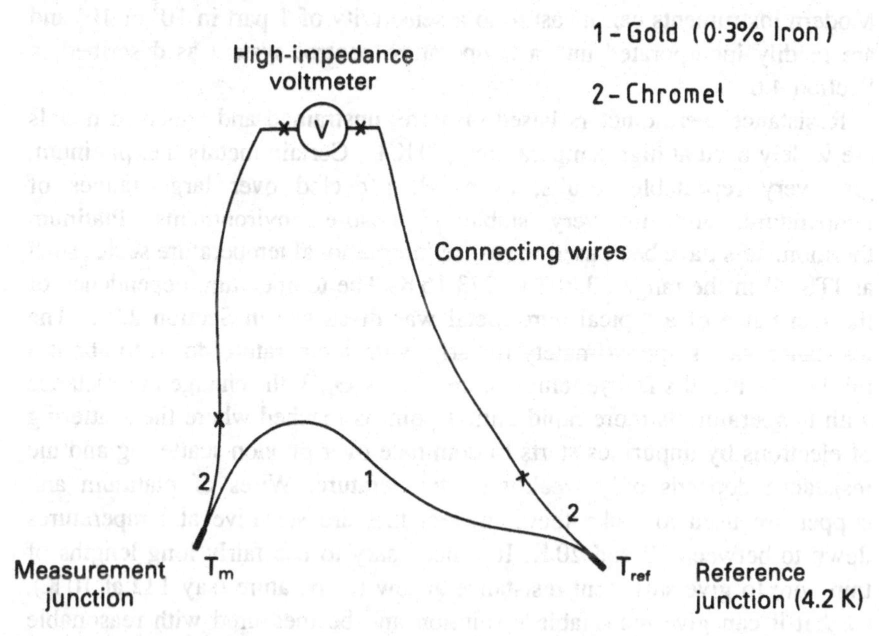

Figure 7.6 *Thermocouple suitable for measuring low temperatures. 1: gold (0.3 per cent iron); 2: Chromel*

Thermocouple junctions can be made very small, and so have a fast response time and minimum thermal load. Their output voltage is readily measured by modern high-impedance digital voltmeters, and the digital output may be fed into a computer, which can be programmed with the calibration equation to give a direct temperature readout. Equation coefficients can be determined from a few fixed calibration points, or from readily available secondary calibration tables. As the temperature reduces, the thermopower of all materials becomes small and eventually tends to zero, which limits the usefulness of thermocouples at low temperatures. Copper—constantan thermocouples have a sensitivity of about 1 μV/K at 4.2 K, compared with about 40 μV/K at room temperature. One combination, AuFe—Chromel (iron component 0.03—0.07 per cent), gives acceptable performance down to about 4.2 K, where vapour pressure techniques take over. The sensitivity is approximately 10 μV/K at 4.2K, though the calibration is very sensitive to the composition of the gold—iron alloy.

7.1.4 Resistance thermometry

One of the most widely used methods of temperature measurement in the range 300—1 K is electrical resistance thermometry. After calibration, resistance thermometers are easy to use, accurate and offer high resolution. Modern instruments can measure to a sensitivity of 1 part in 10^3 or 10^4 and are readily incorporated into a temperature control system as described in Section 4.6.

Resistance thermometers based on pure, unstrained and annealed metals are widely used at high temperatures (70 K +). Certain metals, i.e. platinum, give very repeatable results, even when cycled over large ranges of temperature, and are very stable in hostile environments. Platinum thermometers have been used to realise international temperature scales such as ITS-90 in the range 13.803 to 273.16 K. The temperature dependence of the resistance of a typical pure metal was discussed in Section 2.2.1. The resistance varies approximately linearly, with temperature down to about a third of the metal's Debye temperature. Below $\Theta_D/3$, the change in resistance with temperature is more rapid until a point is reached where the scattering of electrons by impurities starts to dominate over phonon scattering and the resistance depends only weakly on temperature. Wires of platinum and copper are used to make thermometers that are sensitive at temperatures down to between 10 and 20 K. It is necessary to use fairly long lengths of thin wire to give sufficient resistance at low temperature (say 1 Ω at 10 K), so that it can give reasonable resolution and be measured with reasonable accuracy.

Simple circuits (only really suitable for the most non-critical applications) based on metal-sensing elements are shown in Figure 7.7. These circuits are suitable for all the resistance thermometry methods described in this section. A small bias current is passed through the thermometer. It is very important that the bias is kept small, so that the Joule heating within the thermometer does not upset the measurements. The corresponding potential difference V across the sensor is measured, and the resistance is deduced from $R = V/I$, so

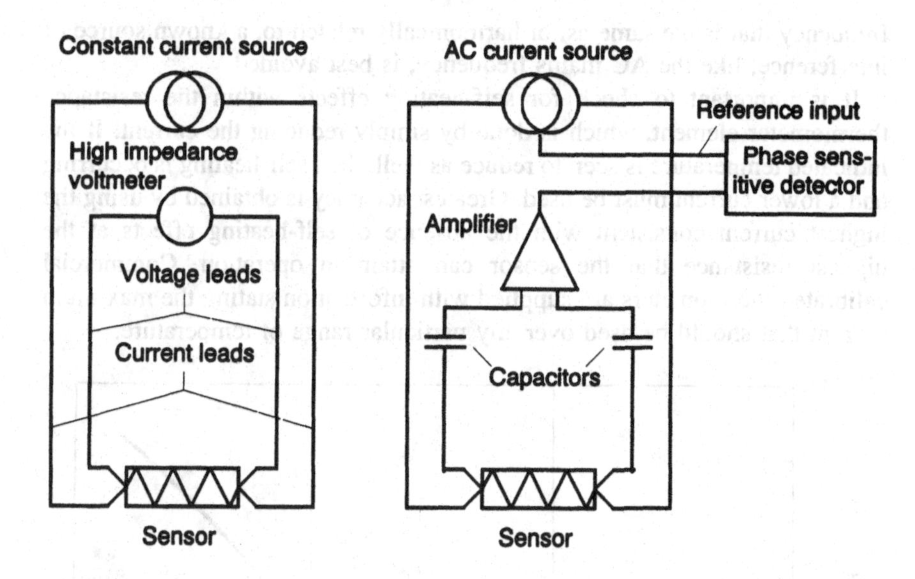

Figure 7.7 *Schematic diagram of simple D.C. and A.C. resistance thermometry systems. In the A.C. system, the capacitors,* C, *isolate any D.C. component that might arise as a result of thermal e.m.f.s*

that the temperature is found from the calibration. The four-terminal technique is preferred, because the connecting leads often have to be long to reach the low-temperature region of the cryostat and are chosen to have a low thermal (and hence electrical) conductivity. With the four-terminal connection, any potential difference across the constant current supply leads is not added to that across the sensor. The current through the voltmeter leads is very small, owing to its high input resistance, and so there is a negligible potential difference across them. Thermal e.m.f.s originating from junctions between different metals and connecting wires subject to large thermal gradients (the Thomson effect, see Section 2.2.3) can be a serious source of error in low-temperature resistance thermometry systems. Alternating techniques may be employed to eliminate these thermal e.m.f.s. With an alternating current delivered to the thermometer resistance element, the potential difference registered by the voltmeter consists of two components: the required one across the sensor which alternates in phase with the current and another which is constant and is the thermal e.m.f. Using suitable measuring equipment, i.e. an alternating-current bridge, it is possible to isolate the alternating and direct components of the e.m.f. Alternating methods also offer the advantage that interference may be rejected by filtering out all except the measurement frequency. A phase-sensitive detector (PSD), or lock-in amplifier, may be used to pick out weak signals oscillating in phase with a reference voltage. The reference signal can be derived directly from the thermometer bias current supply. Using a frequency that is the same as, or harmonically related to, a known source of interference, like the AC mains frequency, is best avoided.

It is important to check for self-heating effects within the resistance thermometer element, which is done by simply reducing the current. If the indicated temperature is seen to reduce as well, then self-heating is occurring and a lower current must be used. Greatest accuracy is obtained by using the highest current consistent with the absence of self-heating effects at the highest resistance that the sensor can attain in operation. Commercial calibrated thermometers are supplied with information stating the maximum current that should be used over any particular range of temperature.

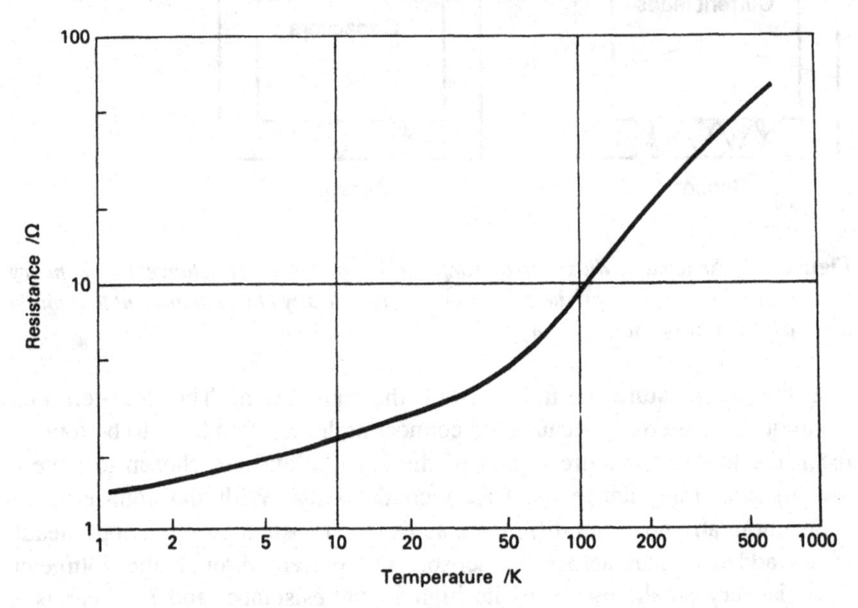

Figure 7.8 *Resistance—temperature characteristics for a typical Rhodium—iron resistance thermometer [Courtesy of Oxford Instruments UK Ltd.]*

Pure-metal resistance thermometers exhibit very little sensitivity at liquid-helium temperatures. However, a number of other materials and devices exhibit sufficiently large resistance changes with temperature in this range. Resistance thermometers based on certain alloys show reasonable resolution at temperatures down to less than 1 K. One example is rhodium doped to about 0.3—0.5 per cent with iron. The temperature dependence of resistance of a typical rhodium-iron thermometer is shown in Figure 7.8. The power-series equation

$$R = a_0 + a_1T + a_2T^2 + a_3T^3 + a_4T^4$$

is found to fit this temperature dependence to within 1 per cent over the range 1 to about 30 K or, with different coefficients, to somewhat above 30

K. The kink in the characteristics around 30 K makes it difficult to get a good fit over the whole temperature range using a single equation. Alloy resistance thermometers become insensitive at low temperatures and even when using sophisticated resistance-measuring techniques it is hard to achieve a resolution better than a few tens of mK at liquid helium temperatures.

In contrast to its behaviour in metals, the resistivity of suitably doped semiconducting materials increases rapidly as the temperature is reduced below about 100 K (see Section 2.2.4). Commercial thermometers based on a germanium element doped with arsenic and encased in a small, helium-filled capsule are widely used for measuring temperatures in the range 0.3 to 100 K. Such thermometers are supplied complete with an accurate calibration curve (see Figure 7.9). Their sensitivity depends on the dopant concentration, which is also used to control the temperature range which the thermometer covers. Typically, the sensitivity of a Ge thermometer is between 100 and 500 Ω/K at 4.2 K. One important advantage of germanium thermometers is the high reproducibility of their measurements, in some cases better than 0.5 mK after thermal cycling between room temperature and liquid-helium temperature. However, this comes at a high price.

Semiconductor *p–n* junction diodes may be used as resistance thermometry elements in the range from over 300 K to liquid-helium temperature. With a constant current flowing in the forward direction, the voltage drop across silicon, germanium and gallium–arsenide diodes is found to increase with decreasing temperature. Ordinary diodes, mass-produced for the electronics industry, are suitable for use as thermometers. The best ones exhibit a smooth variation of the forward voltage drop with temperature over the desired range and are selected by trial and rejection. Purpose-designed gallium–arsenide diode thermometers are commercially available; these are usable at temperatures right down to about 1 K. One of the main problems with all diode thermometers is the high level of bias current needed (>10 μA) to operate them. This is because diodes generate rather a lot of electrical noise at low bias currents. The high bias can give rise to self-heating problems at low-temperatures.

A type of resistance thermometer that offers many advantages for use in the liquid-helium temperature range is based on the traditional, carbon-composition radio resistor. It is found that the resistance of certain types increases rapidly as the temperature is reduced. The temperature dependence is believed to arise from thermal influences on the electron transport across the boundaries between carbon grains in the composite. Figure 7.10 shows the temperature dependence of the resistance of a few examples; it is often found that the resistors having a low room-temperature resistance (up to about 100 Ω) have a greater sensitivity at around 1 K, the higher resistance ones (>100 Ω) being better between 4 and 77 K. Carbon resistors are rarely any use above about 100 K. There is no universal equation for the resistance-

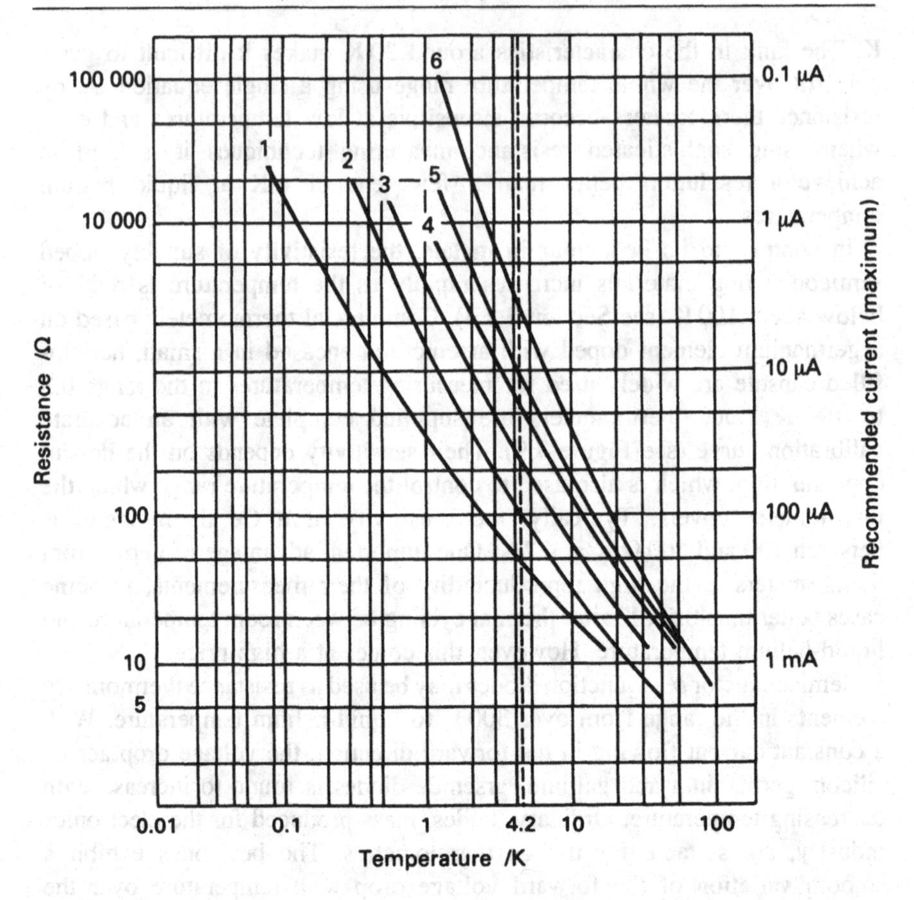

Figure 7.9 *Typical resistance–temperature characteristics for a selection of germanium thermometer elements. The thermometers are characterised by their resistance at 4.2 K. That is, 50, 100, 250, 500, 1000 and 1500 Ω for curves 1–6 respectively. Also shown is the maximum recommended measurement current [Courtesy of Oxford Instruments UK Ltd.]*

temperature characteristics of carbon resistors but an approximate expression

$$T = \sum_{n=-\infty}^{\infty} a_n (\ln R)^n$$

has been found useful in a number of cases. The coefficients a are determined from calibration. Over a restricted range of temperature, say 1–4 K, the reduced equation

$$T = a_{-1}\frac{1}{\ln R} + a_0 + a_1 \ln R$$

can be used with an accuracy of better than 1 per cent and requires a minimum of only three calibration points. Carbon resistance thermometers tend to show small shifts in their calibration upon thermal cycling, and so must be recalibrated when they have been allowed to warm to over 100 K. The carbon composite has a rather low thermal conductivity, and so care must be taken to avoid self-heating effects and to allow the resistor time to reach thermal equilibrium. In spite of their problems, carbon resistors have found widespread favour among low-temperature experimentalists because of their low cost.

Unfortunately with respect to thermometry, carbon resistors have fallen from use in modern electronics and have been replaced by carbon film and metal oxide types, which, sadly, do not seem to make good thermometers. The advent of surface-mounted printed-circuit technology in recent years has led to the development of small chip resistors made of ruthenium oxide, and tests have shown that these make quite good low-temperature thermometers. Their resistance–temperature characteristics below 4.2 K fit

$$R = A \exp(bT^{-1/4})$$

fairly well, which is similar to the behaviour of some types of carbon resistor. The small size of this type of resistor produces a rapid response time, and their stability and reproducibility upon thermal cycling are no worse than those of carbon resistors. Again, the choice of which room-temperature value to use as a low-temperature thermometer depends on the temperature range the resistor is intended to cover; suitable values are in the region of 1 KΩ, which increase in value to about 5 KΩ at 4.2 K.

Carbon–glass resistors, purposefully designed for thermometry in the entire range 1–300 K, have resistance–temperature characteristics similar to those of carbon composition resistors, but they are highly reproducible and are supplied with a calibration. Their sensitivity is typically about 500 Ω/K at 4.2 K. The sensing element consists of pure carbon filaments embedded in a glass matrix, which is sealed in a small metal capsule. Although rather costly, carbon–glass resistors have the advantage that they are little affected by strong magnetic fields, which are present in many low-temperature experiments.

7.1.5 Noise thermometry

Random broadband (white) noise generated in any resistor at a finite temp-

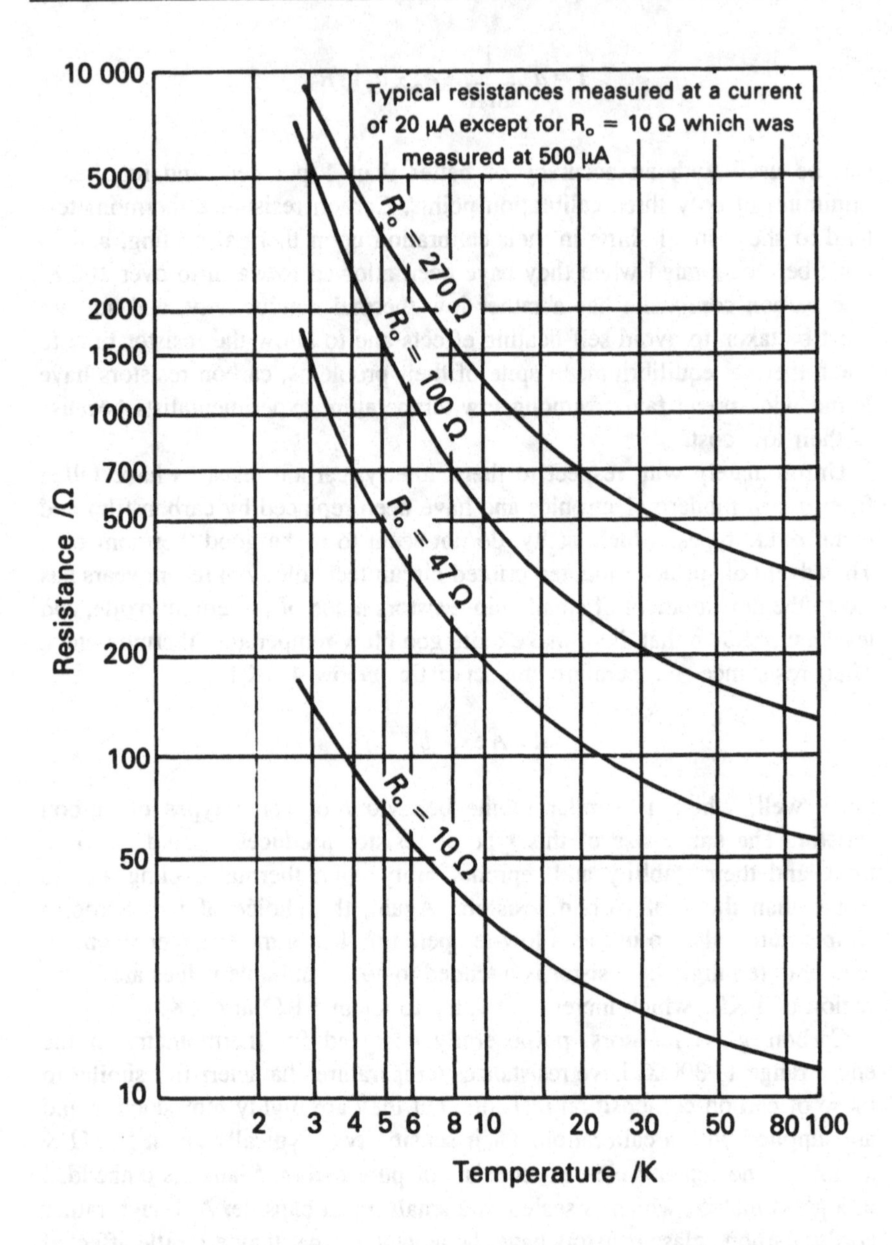

Figure 7.10 *Typical resistance–temperature characteristics for a number of carbon resistors, each having different room temperature values [Courtesy of Oxford Instruments UK Ltd.]*

erature, T, is called Johnson noise. The r.m.s. noise voltage generated in a resistance R at temperature T in a frequency band Δf is given by

$$\bar{v} = (4kTR\,\Delta f)^{1/2}.$$

The noise voltage is rather small, typically of the order of microvolts, but, given a low noise amplifier of well-defined gain and bandwidth, $\bar{v}$ can be measured and the temperature calculated using the above equation. Noise thermometry is analogous to gas thermometry; it is due to the temperature-dependent, random thermal motions of electrons instead of gas atoms.

The main problem with noise thermometry is the small size of the noise voltage at low temperatures. For example, in a resistor of 100 kΩ at 4.2 K, the r.m.s. noise voltage in a bandwidth of 1 MHz is only about 1.5 μV. It would be a very good amplifier indeed that did not have an inherent noise level at least comparable to this. Another problem involves the determination of the bandwidth, Δf, of the measuring apparatus, which gets harder as attempts are made to increase it to obtain a larger noise voltage. A clever technique (Kamper *et al.*, 1971) that overcomes both of these problems and allows noise thermometry to be used down to very low temperatures involves connecting the resistor across a SQUID. By applying a small bias voltage, the SQUID can be made to operate as a voltage-controlled oscillator in the kilohertz range. The noise voltage across the resistor modulates the bias and hence the SQUID frequency. Frequency can be measured very accurately. The mean square deviation of a number of measurements is given by

$$\sigma^2 = \frac{8ke^2RT}{h^2\tau},$$

where τ is the time taken to make an individual frequency measurement. Using this technique, noise thermometry can give a resolution better than 0.1 per cent at 1 K.

7.1.6 Mounting thermometers in low-temperature apparatus

Very good thermal contact between the thermometer and the sample is vital to give negligible temperature errors and the quickest response time. Commercial resistance thermometer elements are often encapsulated within a small cylindrical package, about 3 mm in diameter and 10–20 mm long. These can be inserted into a close-fitting hole drilled in a copper block and held in place by a coat of varnish. The block may be an integral part of the apparatus, such as the sample holder, or it may be a separate component placed in close thermal contact with a part of the cryostat. Close thermal contact is ensured by tightly screwing or clamping the block down; the mating faces should form a clean electrical contact. On a microscopic scale, the surfaces of metals are rough. When two pieces of metal are put together, it is only the 'high points' that actually make contact. The effective contact area, and therefore the thermal conductance, is increased by pressing the surfaces together harder. This causes the microscopic ridges to deform and fill the gaps.

It is difficult to get a very good thermal contact to carbon resistors because both the carbon composite and the encapsulation material have poor thermal conductivity. Some improvement may be gained by carefully scraping away the encapsulation to expose the carbon composite. A thin layer of electrically insulating varnish is painted onto the exposed composite before mounting.

Electrical leads should be tightly thermally anchored close to the thermometer. This improves the thermal contact to the thermometer element and, most importantly, prevents any heat that is conducted down the wires from reaching the thermometer. The wires should be wound round the thermometer mounting point a number of times and held in place with varnish or glue. To limit further the conducted heat reaching the thermometer, it is wise to thermally anchor the thermometer leads at various points on their way down the cryostat.

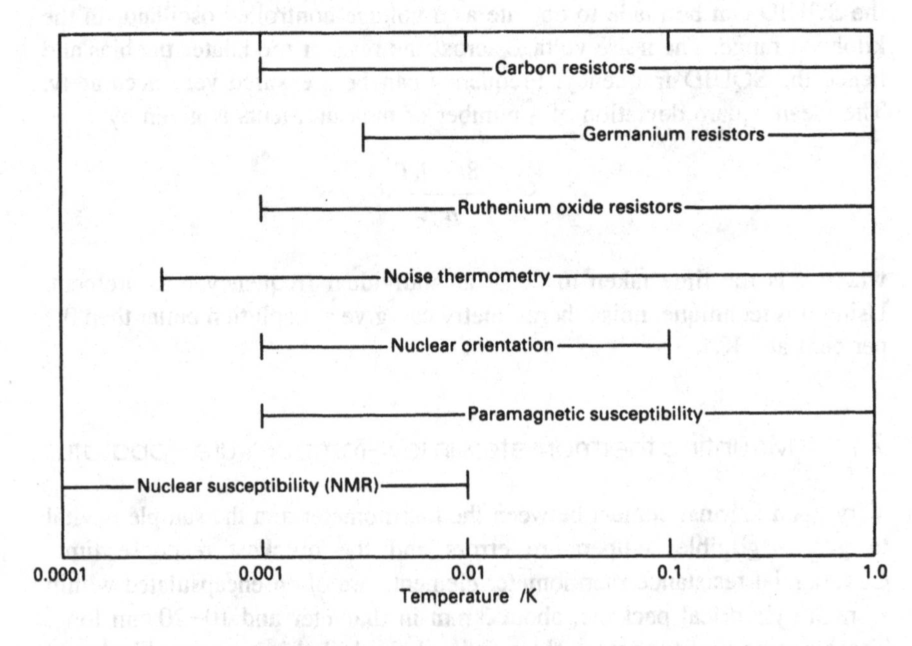

Figure 7.11 *Types of thermometer suitable for use below 1 K, indicating the approximate temperature range over which they are used*

7.2 Thermometry below 1 K

Figure 7.11 shows some of the types of thermometer used below 1 K. The same problems of thermal contact and internal heating encountered at higher temperatures are experienced in the millikelvin temperature range, but are accentuated as heat capacities and refrigerator cooling powers plummet. Picowatt (10^{-12} watt) power dissipation is all that can be allowed before self-heating within the thermometer element becomes intolerable. Some of the thermometry techniques used at temperatures above 1 K can be extended to even lower temperatures. Germanium, carbon and ruthenium oxide resistance thermometers are all used at millikelvin temperatures. In spite of the difficulties involved with operating at very low power levels, resistance thermometers are the mainstay of millikelvin laboratories, because compared with other methods they are so convenient. However, a means of calibrating the resistors below the minimum observable helium vapour pressure is required. Noise thermometry, described in the Section 7.1.5, is useful for this purpose down to about 1 mK, but is a very specialised technique at such low temperatures.

7.2.1 Magnetic susceptibility thermometry

An ideal paramagnetic material is one in which there are no interactions between the magnetic dipoles. At low magnetic field strength, the magnetic susceptibility of an ideal paramagnetic material follows Curie's law,

$$\chi = \mu_0 \frac{M}{B} = \frac{C}{T},$$

where B is the field acting on the dipoles and M the magnetisation (see Chapter 6). The Curie constant C may be calculated from first principles, so measurement of the ideal susceptibility yields absolute temperature directly. Unfortunately, at a particular value of applied field, as the temperature falls, the magnetisation increases and a point is reached where the magnetisation tends to saturate. To extend the measurement range below this point, the applied magnetic field may be reduced. The minimum value of the applied field which may be used is that which is comparable to the internal fields. This determines the minimum temperature that may be measured using paramagnetic susceptibility. The salt cerium magnesium nitrate (CMN), used for cooling by adiabatic demagnetisation, is used in paramagnetic thermometry between about 2 mK and 1 K because it has relatively low internal fields. Using dilute magnetic media in which the magnetic atoms are well separated (for example, CLMN with about 5 per cent cerium to lanthanum) reduces the problem of internal fields to some extent, but the susceptibility is smaller and, consequently, harder to measure accurately.

Figure 7.12 (top) shows schematically a simple arrangement for magnetic thermometry. The paramagnetic sample is placed within two coils, which make a transformer, and the mutual inductance between them is measured using A.C. techniques. The mutual inductance is given in terms of the

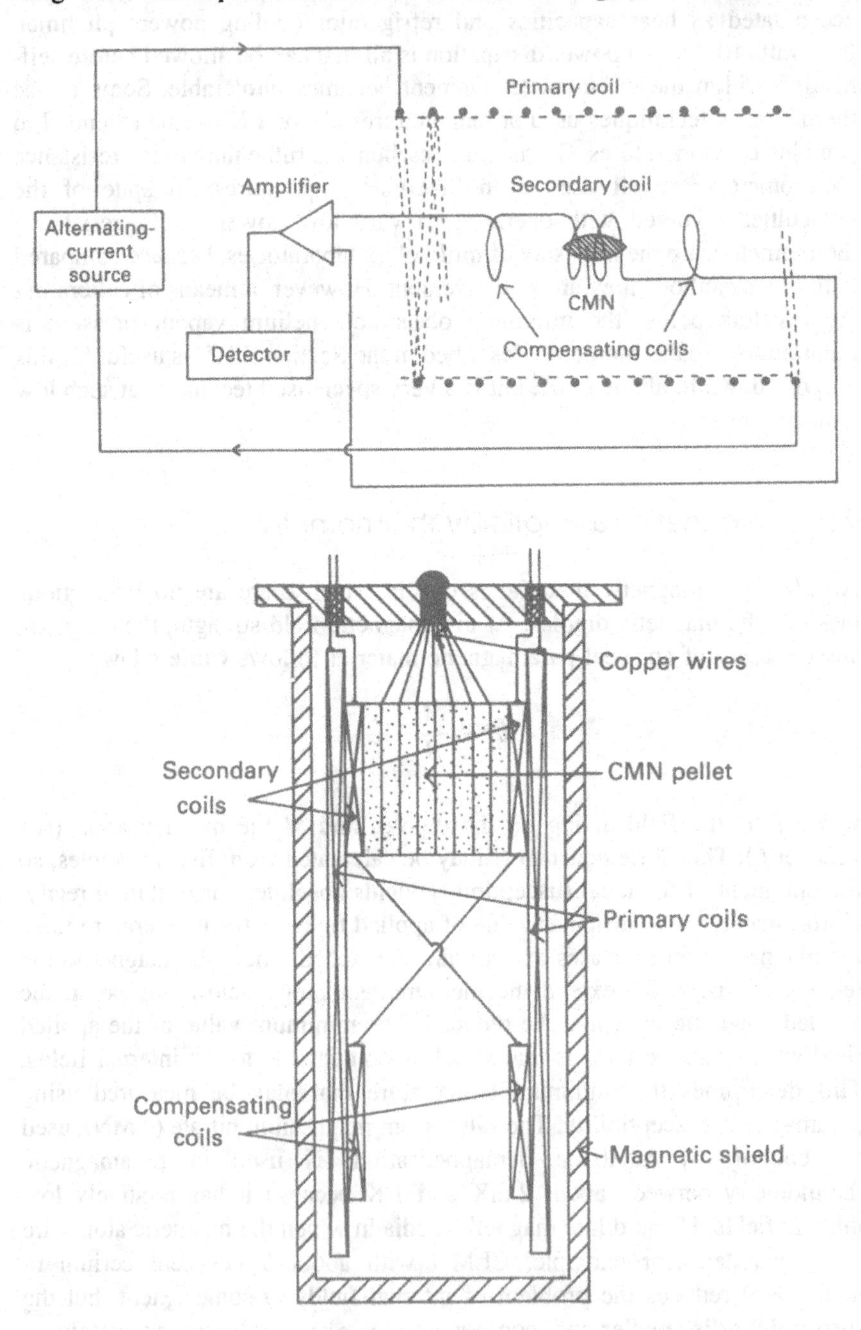

Figure 7.12 (top) *Schematic diagram of CMN magnetic susceptibility thermometer system.* (bottom) *An experimental magnetic thermometry cell*

susceptibility by

$$M = M_0(1 + f\chi),$$

where M_0 is the mutual inductance in the absence of the sample and $f < 1$ is the filling factor constant which depends on how much of the secondary coil's volume is occupied. For a coil embedded in an infinite block of the paramagnetic, $f = 1$. Compensating coils are arranged to balance out the constant term, M_0; they are wound in opposition to the coil surrounding the sample. The mutual inductance is then given by

$$M = M_0 f\chi.$$

The shape of the sample is important because a demagnetising field, B_d, is set up within it. In ellipsoidally shaped samples, B_d is proportional to the magnetisation and given by

$$B_d = \mu_0 \varepsilon M.$$

For a spherical sample, $\varepsilon = 1/3$; for an ellipsoid of revolution, it is dependent only on the ratio of length to width. In addition to the demagnetising field, there is the internal field arising from the interactions between the dipoles. If the dipoles are on a cubic lattice, this is also proportional to the magnetisation and given by

$$B_i = \mu_0 M/3.$$

The field B acting on the dipoles is related to the applied field B_a by

$$B = B_a - B_d + B_i.$$

Combining these gives the susceptibility in terms of the applied field

$$\chi = \mu_0 \frac{M}{B_a} = \chi_0 + C\left[T - \left(\frac{1}{3} - \varepsilon\right)C\right]^{-1},$$

where χ_0 allows for any possible temperature-independent susceptibility component.

Figure 7.12 (bottom) shows a typical experimental thermometer head. It may be about 1 cm in diameter and a few centimetres long. Good thermal contact between the CMN pellet and the copper mounting plate is ensured by embedding many tiny copper wires in the pellet and soldering their ends to the plate. If care is taken not to heat the sample by the A.C. field, and also to keep stray magnetic fields away from the thermometer head, susceptibility thermometry is capable of very high sensitivity and accuracy.

To measure temperatures much below a millikelvin, a nuclear paramagnetic material must be used because its internal field is much smaller. However, the method used to measure the static magnetic susceptibility of electronic paramagnetic materials lacks sufficient sensitivity

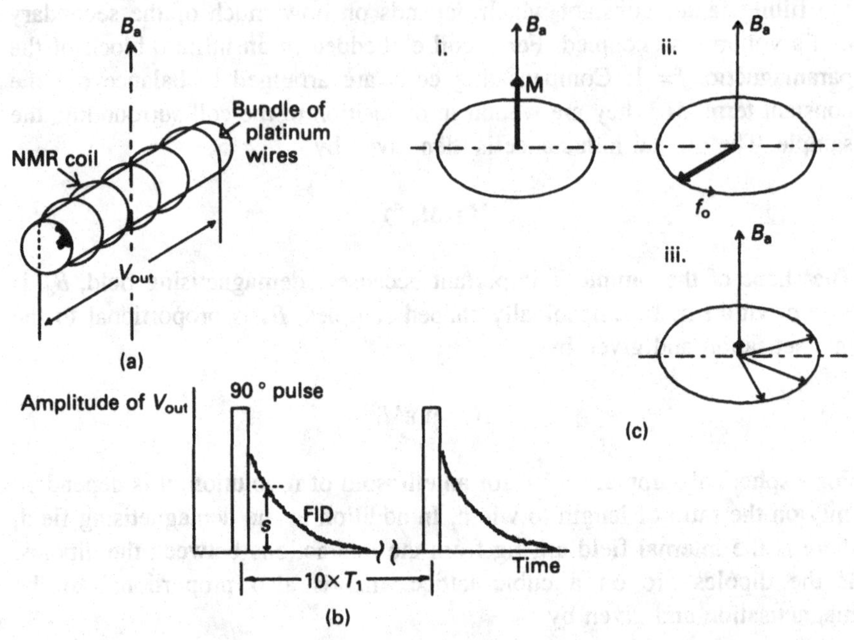

Figure 7.13 *(a) Geometry of an NMR thermometer probe. (b) Amplitude of voltage signal across an NMR coil as a function of time. (c) Evolution of the magnetisation during an NMR experiment*

for use with nuclear materials. The nuclear moment is a thousand times less than the electronic moment, and the sensitivity of the above method is proportional to μ^2, i.e. a million times less. For this reason, susceptibility is determined using the more sensitive and selective technique of nuclear magnetic resonance (NMR) (see Figure 7.13). If a nuclear paramagnetic sample is placed in a magnetic field then the nuclear dipoles precess about the field at a characteristic frequency, $f_0 = \gamma B/2\pi$ where γ is the magnetogyric ratio, just as a mechanical gyroscope precesses about the earth's gravitational field. Typically, f_0 is in the radiofrequency range $f_0 = 1$–40 MHz in fields of order 1 T. If the sample is placed in a coil oriented at right angles to the applied field (Figure 7.13a), and the coil is energised by a strong pulse of radiofrequency current with a frequency equal to the precession frequency, then the magnetisation, M, is tipped away from the direction of the applied magnetic field towards the axis of the coil as shown in Figure 7.13c. If the pulse strength and duration are chosen correctly, the magnetisation is tipped into the axis of the coil, i.e. at right angles to the applied field. Such a pulse is called a *90-degree* pulse. Following the pulse, the dipoles continue to precess about the applied field, inducing a small voltage in the coil; this is the NMR signal. In a short time the dipoles get out of phase, owing to slight variations in their individual precession speeds and the signal decays exponentially with characteristic time T_2^*. This decay of the NMR signal is called free induction decay (FID) (Figure 7.13b). Over a further period of time, the magnetisation forms up parallel to the applied field again. The characteristic time for this to happen is called the spin-lattice relaxation time. Platinum is an ideal choice for NMR thermometry, firstly, because its relatively long value of $T_2^* \sim 1$ ms, compared with that for other metals, allows easy measurement of the FID, and secondly, because its short value of T_1 ($\sim 29.6/T$ ms) allows for relatively rapid repeat measurements, for it is necessary to wait a time at least $10 \times T_1$ for the magnetisation to be fully parallel to B_a. In common with other metals, platinum has relatively high thermal conductivity and reaches internal equilibrium quickly. The amplitude, s, of the free induction decay signal is proportional to the initial magnetisation of the sample, and hence the susceptibility. If the FID amplitude is known at some fixed temperature, then other temperatures can be inferred from the relative signal heights. Metal nuclear magnetic thermometry is the method used at the lowest temperatures currently achieved by nuclear demagnetisation.

The main drawback of nuclear magnetic thermometry, as the temperature is reduced, is eddy current heating of the sample by the radiofrequency fields. There is also a loss of sensitivity owing to the skin effect, which attenuates high-frequency fields as they penetrate into a conductor. To alleviate these problems, the samples are fabricated from bundles of thin wires or powders.

In principle, both of the above techniques can be used for primary thermometry. However, measurement of the absolute value of the susceptibility is difficult. Magnetic thermometers are generally calibrated at the high end of their temperature range using fixed points and the calibration extrapolated to lower temperatures, assuming that Curie's law is obeyed. The temperature measurement obtained in this way is called the magnetic temperature and its deviation from the true temperature increases as the temperature reduces.

7.2.2 ^{3}He melting curve thermometry

Melting curve thermometry (MCT) is based on the properties of the ^{3}He melting curve (Figure 5.7). To each temperature on the melting curve there corresponds a unique value of the pressure. Absolute values of pressure can be measured to a resolution of better than 10^{-3} Pa in a capacitance cell, similar to that shown in Figure 7.14. Thermal contact between the helium

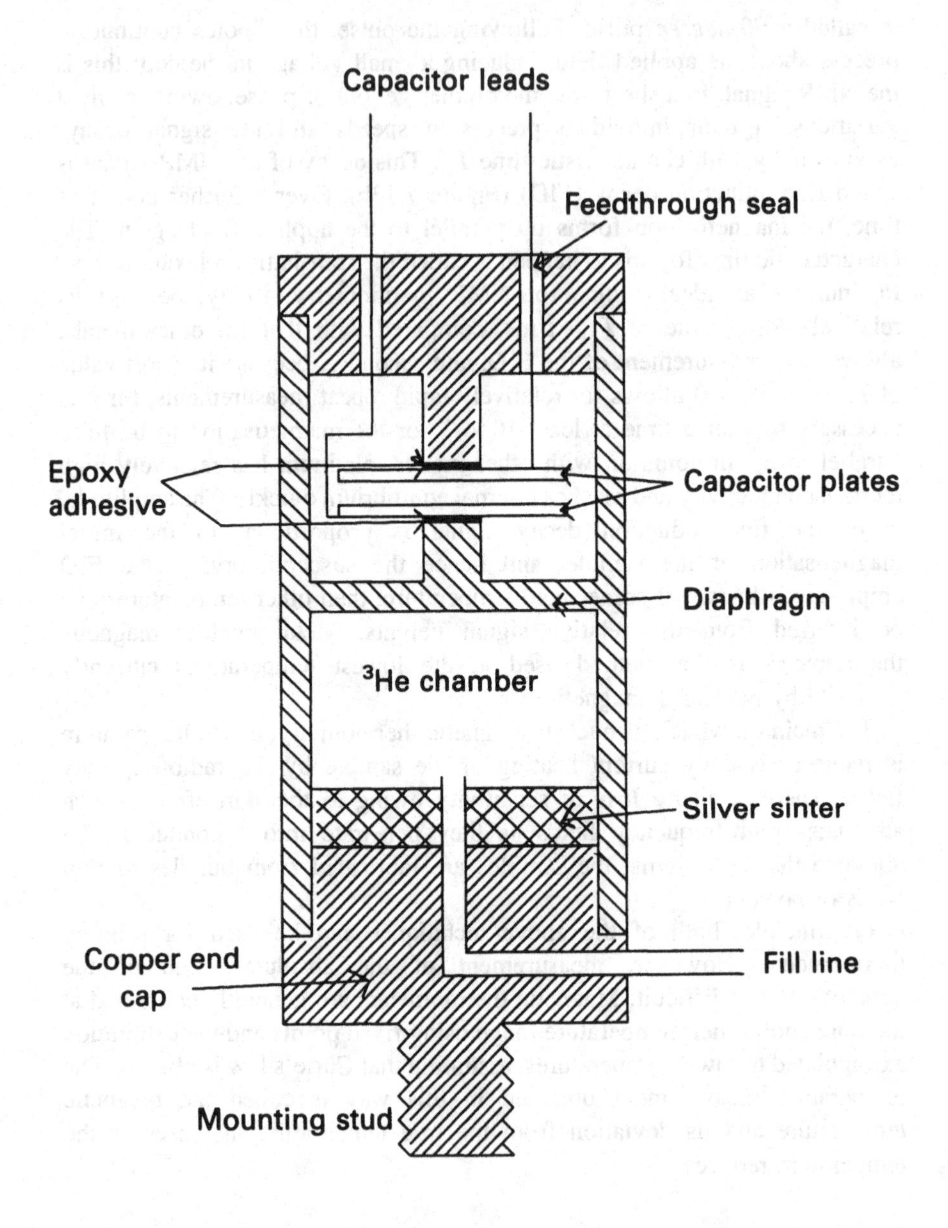

Figure 7.14 *Pressure cell suitable for ^{3}He melting curve thermometry*

and the end cap–mounting stud assembly is aided by the silver sinter, a 'sponge' made of tightly compressed silver particles, which has a large surface area to volume ratio. Determinations of $p(T)$ along the melting curve, based on a combination of theory and experimental data, have been carried out. In practice slight corrections to the calculated relationship may be necessary. These can be determined by calibration at fixed points such as the ^3He superfluid transition. The relationship between pressure and temperature along the melting curve is found to follow the power-series form (Greywall 1986)

$$p = p_A + \sum_{n=-3}^{n=5} a_n T^n$$

which is accurate to within 1 μK for T in the range 1 to 250 mK. Here p_A is the normal–superfluid (A-phase) transition on the melting curve.

7.2.3 Nuclear orientation thermometry

Nuclear orientation thermometry is capable of providing absolute measurements in the millikelvin range and is useful as a standard against which other types of thermometer, such as magnetic susceptibility thermometers, can be calibrated. The sensing element is just a small piece of weakly radioactive metal which can be clamped or soldered directly to the apparatus to obviate problems of thermal contact and self-heating. A great advantage is that no direct connection to the sensor is required; the radioactive decay products are detected outside the cryostat, which is transparent to them.

The radioactive (gamma) decay of certain nuclei (e.g. ^{60}Co, Figure 7.15a) is spatially anisotropic. The angular distribution of the gamma ray emission depends on both the nuclear decay properties and the state of polarization of the nucleus. The decay properties are normally temperature independent, but if the nucleus is placed in a magnetic field then the degree of polarisation increases as the temperature is reduced. In a magnetic field, the energy levels of a nucleus of spin I are split into $2I + 1$ hyperfine levels separated by $\Delta E = g\mu_N B$, where μ_N is the nuclear magneton and g the nuclear g-factor. The magnetic field, B, is provided by adding, substitutionally, the radioactive nuclei to a ferromagnetic host such as ^{59}Co. Typically, B is of order 10 T, giving an energy level separation (in temperature terms) of $\Delta E/k = 3.1$ mK. The relative occupation of each of the hyperfine levels is given by the Boltzmann equation and the degree of polarization is proportional to the factor $kT/\Delta E$. Each hyperfine level has its own anisotropic gamma-decay properties and the distribution $W(\theta, T)$ of gamma rays emitted by the sample is the sum of these weighted by the relative occupations of the levels.

Figure 7.15b shows the arrangement for nuclear orientation thermometry based on a ^{60}Co source. The sample is usually in the form of a small needle and the emission angle, θ, is measured relative to the axis of the needle (the crystallographic c-axis in this case). A scintillation detector outside the cryostat is used to count the gamma rays at different angles. At high temperatures ($kT/\Delta E \rightarrow \infty$) the emission is isotropic, while at low temperatures it is peaked strongly at right angles to the axis of the needle. The appropriate nuclear hyperfine constants can be measured and so, in theory, it is possible to compute with some accuracy the dependence of the distribution $W(\theta, T)$ on T. Figure 7.15c shows the results of such a calculation for ^{60}Co in a ^{59}Co host lattice. By comparing these curves with the experimental data, the temperature T may be deduced. The sensitivity is greatest at around 10 mK and the minimum and maximum temperatures that can be measured are about 2 mK and 40 mK respectively. Counting error is proportional to $N^{1/2}$, where N is the total number of counts recorded, and so a longer count will result in greater accuracy. Accuracies of a few per cent at around 5 mK are possible before the counting delay becomes unacceptable.

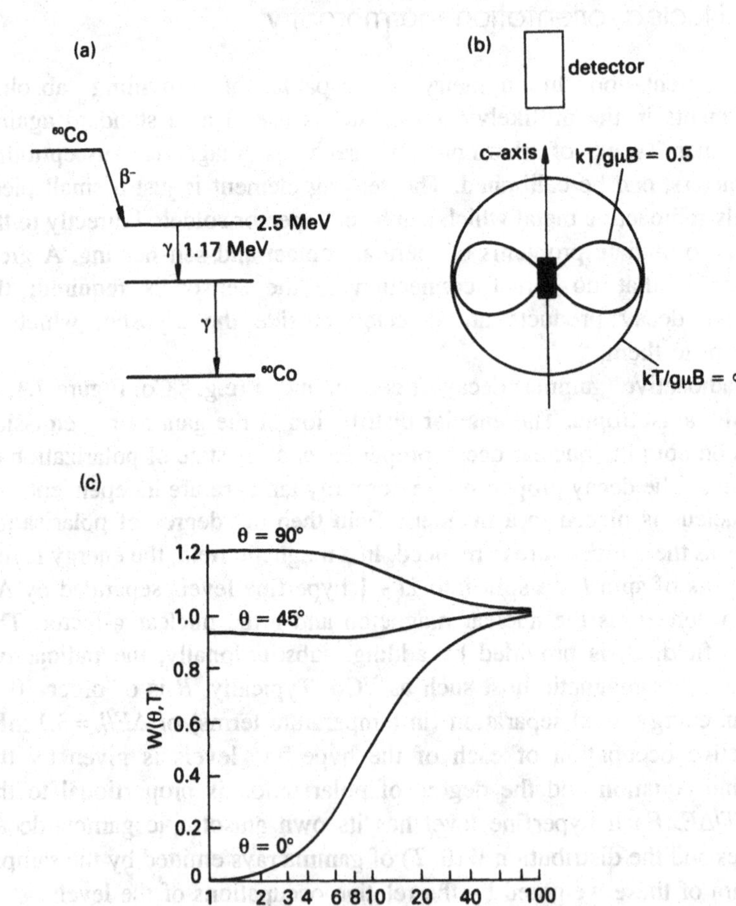

Figure 7.15 *(a) Radioactive decay scheme for* ^{60}*Co. (b) Schematic diagram of nuclear orientation thermometer, the cobalt crystal is inside the cryostat. (c) Theoretical dependence of* W(θ, T) *on temperature [Betts, 1989]*

7.3 Temperature measurement in the presence of magnetic fields

Many low-temperature experiments are carried out in the presence of strong magnetic fields. With the exception of the gas and vapour pressure thermometers, all the methods discussed in this chapter are affected to some extent by magnetic fields. Some require the application of a known field to work at all and can, therefore, be used only if this is the same field as is used in the experiment. Metal, alloy, semiconductor and carbon resistance thermometers all display magnetoresistance (a change of resistance in the presence of a magnetic field). The worst affected are semiconductor (i.e. germanium) thermometers. Metal and alloy resistance thermometers show comparatively little magnetoresistance, with carbon resistors falling somewhere in between. The magnetoresistance increases as the temperature is reduced. For a typical carbon resistor in a field of 1 T, it can amount to about 0.1 per cent at about 10 K, increasing to about 2 per cent at 1 K. If it is possible to do so then the resistance thermometer should be positioned in a zero- or low-field region and thermally linked to the sample by a high-thermal-conductivity wire or rod (i.e. highly pure copper). In spite of their expense, carbon-glass thermometers may be a good choice because of their very small magnetoresistance, about 2 per cent in a field of 10 T. Thermocouples may be used to measure temperature in the presence of a magnetic field, provided that the connecting wires are at constant temperature in a field gradient.

At low temperatures, the dielectric constant, ε, of some materials is temperature-dependent. It is also very insensitive to magnetic field. In glass (SiO_2), ε has an approximate $\ln (T^{-1})$ dependence below 100 mK. The sensitivity of ε to changes in temperature is increased by using glass with a large OH^- content, approximately a thousand parts per million. Capacitance thermometers are made by evaporating metal electrodes onto the opposite faces of a very thin glass film. The capacitance is measured by using an A.C. bridge. Typically, the variation of capacitance with magnetic field strength is less than 0.1 per cent up to 20 T. Capacitance thermometers are, therefore, a very good choice for use in strong magnetic fields. They also operate with negligible self-heating. However, they do not give very reproducible results upon thermal cycling, and do tend to deteriorate with usage.

Bibliography

Betts, D. S. (1989). *An Introduction to Millikelvin Technology* (Cambridge: Cambridge University Press)

Greywall, D. S. (1986). *Phys. Rev.*, **B33**, 7520

Kamper, R. A., Siegworth, J. D., Radebaugh, R. and Zimmerman, J. E. (1971). *IEEE*

Proc., **59**, 1368
Quinn, T. J. (1990). *Temperature*, second edition (London: Academic Press)
Roberts, T. R. and Sydoriak, S. G. (1957). *Phys. Rev.*, **106**, 175
Rusby, R. L. and Durieux, M. (1984). *Cryogenics*, **24**, 363
Steinberg, V. and Ahlers, G. (1983). *J. Low Temp. Phys.*, **53**, 255

8

Experimental techniques, hints and tips

Many of the experimental techniques originally devised at room temperature now have been extended to low temperatures. There are two main reasons for taking the trouble to do this. Firstly, additional information regarding the temperature dependence of physical properties can be obtained. The new information can be compared with theoretical models which may, in some cases, be simplified in the absence of strong thermal disordering. Secondly, certain phenomena such as, for example, superfluidity in liquid ^{4}He (see Chapter 3), occur only at low temperatures. Extending an experimental technique into the low-temperature regime introduces a whole new set of problems to overcome. In this chapter just a few of these problems and their practical solutions will be considered.

8.1 Vacuum technique

Vacuum technique is an important element of low-temperature experimentation. A high vacuum is required to achieve good thermal isolation in cryogenic apparatus. Vacuum pumps are also used to reduce the vapour pressure over liquid cryogens and hence their boiling temperature (see Section 4.5). A basic understanding of vacuum technique is, therefore, an essential requirement for a low-temperature experimentalist.

A number of different types of vacuum pump are available on the market. Which one to select for a particular application depends on the pumping speed and ultimate vacuum pressure requirements. The pumping speed is a measure of the volume throughput per unit time of the pump. Pump manufacturers normally express the pumping speed in litres per minute at a certain pressure (or over a range of pressures). The pumping speed reduces rapidly to zero as the base pressure or ultimate vacuum pressure of the pump is approached.

8.1.1 Rotary vacuum pumps

Mechanical rotary vacuum pumps are capable of attaining pressures between atmospheric and about 10^{-2} Pa (approximately 10^{-7} atmospheres). They are available in various sizes, having volume pumping speeds between 10 and more than 10 000 l per minute. A typical rotary pump consists of a rotating cylinder within a cylindrical container (Figure 8.1). The axes of the two cylinders are offset and the rotating cylinder has spring-loaded vanes which press against the surrounding walls of the fixed container to form a gas-tight seal. The rotating cylinder is driven by an electric motor, either directly or via a belt. Moving parts are both lubricated and sealed by a special low-vapour-pressure oil. The gas, at low pressure, entering the volume is trapped by the vanes as they sweep around and compressed until its pressure is sufficient to lift the exhaust flap, whereupon it is expelled from the pump. Theoretically the minimum pressure that can be reached is simply related to the geometry of the swept volume and the exhaust pressure. When the pressure built up in the pump is insufficient to lift the exhaust flap, the gas is carried back round to the input port again.

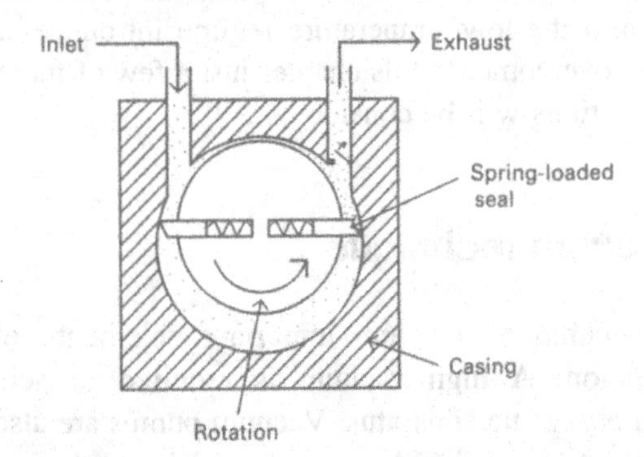

Figure 8.1 *Cut away view of a rotary vacuum pump. The rotating cylinder is driven by an electric motor. The sliding vanes are both sealed and lubricated by a special low-vapour-pressure oil*

A higher vacuum can be obtained with a two-stage rotary pump. The first-stage exhaust is connected directly to a second, smaller stage, thus reducing the pressure that needs to be built up in the first stage in order to exhaust the gas.

Any good one- or two-stage rotary pump can achieve a sufficiently low base pressure to lower the temperature significantly when pumping a helium bath. When choosing a pump for this purpose it is more important to consider its speed: The pumping speed determines the maximum cooling power and hence the ultimate base temperature. Some compromise is necessary because faster pumps are physically larger, more expensive and consume more power.

It is not unusual for rotary pumps to have small leaks in their casing, which allow small amounts of air to mix with the pumped gases. Normally this does not matter in cryogenic applications. However, pumps are made that have been carefully sealed to avoid loss or contamination of pumped helium gas. These should be employed when pumping the valuable isotope ^{3}He.

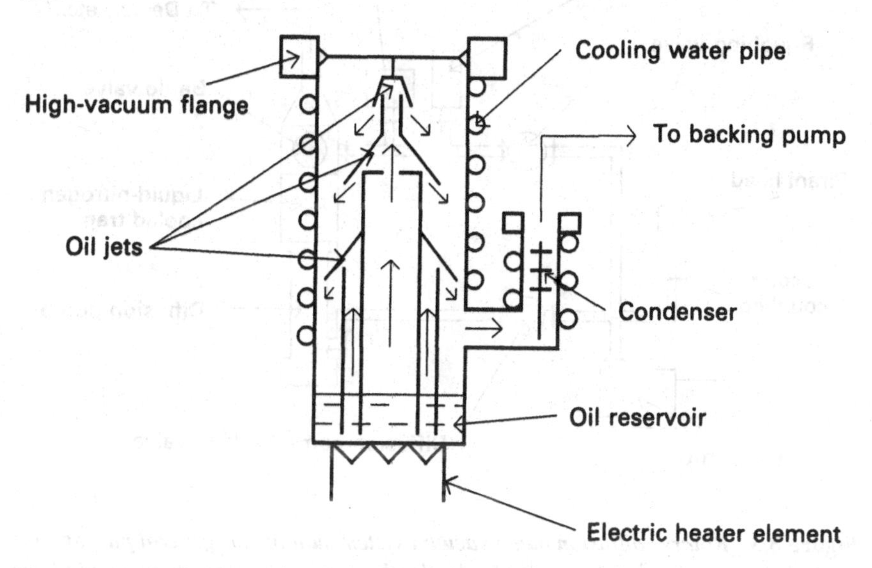

Figure 8.2 *Internal view of an oil vapour diffusion pump*

8.1.2 Vapour diffusion pumps

The base pressure achieved by a rotary pump alone is often insufficient to give adequate thermal isolation, for example in Dewar vessels. Vapour diffusion pumps are capable of attaining pressures down to about 10^{-6} Pa with volume pumping speeds of between 50 and 50 000 l per second when used in conjunction with a suitable rotary pump. Early vapour pumps contained mercury, but for safety reasons modern pumps are designed to operate with heavy synthetic, silicone or mineral-based oils. Figure 8.2 shows a cross-sectional view of a typical vapour diffusion pump. The oil reservoir at the bottom of the pump is electrically heated and the oil vapour rises up inside the pump. The walls of the pump are cooled by water or forced air flow. When the oil vapour meets the walls it cools and falls back down the pump, eventually condensing and running back into the reservoir. The inside of the pump is shaped so as to maintain a circulatory flow of oil vapour.

Gas molecules entering the top of the pump become trapped in the high-velocity stream of heavy oil molecules and are dragged along with them. A similar effect might be experienced by a pedestrian trying to walk into a football ground as the crowd emerges at the end of a match. The gas pressure at the base of the pump is prevented from rising significantly by connecting the lower end to a rotary, backing, pump.

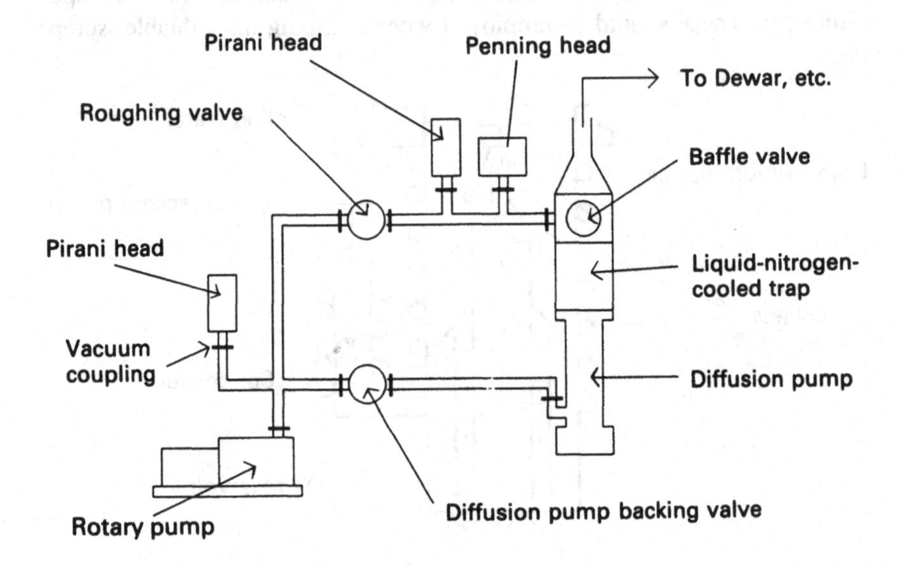

Figure 8.3 *Rotary–diffusion pump vacuum system suitable for general purpose use in the low-temperature laboratory. Typically, the two-stage rotary pump would have a speed of about 100 l/min and the diffusion pump a bore of about 50 mm. The liquid-nitrogen-cooled trap is not essential, except where it is important to minimise backstreaming and to maximise the pumping speed for water vapour*

It is essential that diffusion pumps are operated at an input pressure of less than about 10 Pa and that the output is backed with a rotary pump. If these normal operating conditions are not maintained then the diffusion pump will stall and the oil may become damaged. Figure 8.3 shows a simple vacuum system based on rotary and diffusion pumps suitable for multipurpose high-vacuum applications in a low-temperature laboratory. The system is operated as follows: With all of the valves except the backing valve closed, the rotary pump is switched on to evacuate the diffusion pump. The power to the diffusion pump heater is then turned on. After about fifteen minutes, when the oil has reached operating temperature, the liquid-nitrogen trap (if fitted) may be filled. At this stage the backing valve is closed, the roughing valve opened and the workpiece pumped out to a 'rough' vacuum of < 10 Pa. Only then should the diffusion pump be connected to the workpiece by closing the roughing valve,and opening the backing and baffle valves in that order.

One problem that can be experienced with diffusion pumps is the backstreaming of small amounts of oil vapour into the vacuum system. This can lead to the damage of delicate items, such as dilution refrigerators, which may become blocked with frozen oil. Backstreaming can be minimised at the expense of pumping speed by including water- or liquid-nitrogen-cooled baffles or traps above the pump to condense and catch the oil vapour.

8.1.3 Turbomolecular pumps

An alternative to the diffusion pump for attaining low pressures is the turbomolecular pump. In a turbomolecular pump the inlet gas is compressed by a set of rotating turbine blades, similar, in principle, to a jet engine. There is no oil to backstream and the ultimate vacuum that can be reached is about 10^{-8} Pa. Pumping speeds between about 100 and 1000 l per second are possible. The mode of operation is similar to that of the diffusion pump. It must be backed by a rotary pump and run at an inlet pressure below about 100 Pa. The disadvantages of turbomolecular pumps are that they are rather expensive, delicate and create too much vibration for some cryogenic applications. For these reasons they only tend to be used where the requirements for the lowest ultimate vacuum and no backstreaming are paramount.

8.1.4 Cryopumps

A gas molecule coming into contact with a solid surface experiences a van der Waals force. If the surface is cooled, the molecule's thermal energy may not be sufficient for it to escape from the attracting force, and the molecule is adsorbed onto the surface. In a cryopump, a material of large specific surface area is cooled to cryogenic temperatures to adsorb small traces of gas. Fairly large pumping rates, up to 5000 l per second, and extremely low base pressures (better than 10^{-10} Pa) are possible. This is the principle behind the sorption pump in the ^{3}He cryostat described in Section 5.1. The disadvantage of the cryopump is that, once the surface is saturated, it cannot pump further until it is warmed and the liberated gas is pumped away by a rotary or other pump. It is, therefore, a 'single-shot' device.

Most low-temperature apparatus is not designed with cryopumping in mind. However, many low-temperature experimenters unwittingly benefit from cryopumping. When a Dewar is filled with liquid helium, its inner walls cool to 4.2 K and adsorb any remaining traces of gas in the vacuum space. This enhances the vacuum produced by rotary and diffusion pumping and improves the thermal isolation.

8.1.5 Mechanical booster pumps

A mechanical booster or Roots pump, backed by a rotary pump, may be used to achieve a very high pumping speed, up to about 100 000 l per minute, at moderately low pressures, 10 to 10^4 Pa, e.g. for pumping a ^{4}He cryostat. The disadvantage with the Roots pump is that, because it has to be manufactured to within very close mechanical tolerances, it is much more expensive than rotary or diffusion pumps.

8.1.6 Pressure measurement

For measuring pressures above about 10 Pa (0.1 mbar), manometers are accurate but rather cumbersome. Mercury or oil manometers are widely used for measuring the vapour pressure over a pumped helium bath in thermometry. The two most common pressure indicators in high-vacuum systems are Pirani and Penning gauges. These are readily available from a number of manufacturers, are straightforward to install and operate and cover a wide range of pressures. However, they are not particularly accurate.

The Pirani gauge exploits the fact that the rate of heat conduction between two surfaces by gas molecules is proportional to the pressure, provided that the mean free path of the molecules is larger than the separation of the surfaces. The gauge consists of an electrically heated wire surrounded but not touched by a sleeve. This arrangement is mounted within the vacuum chamber or pumping line. The temperature of the wire is determined from its resistance and hence the rate of heat loss by conduction through the gas to the sleeve is computed. The typical range of a Pirani gauge is from 10^4 to 0.1 Pa (100 to 10^{-3} mbar), which is ideal for measuring the range of pressures attainable using a rotary pump.

The Penning or ionization gauge consists of two electrodes separated by a few millimetres, across which a high potential difference (> 1000 V) is applied. Electrons emitted from the negative electrode are accelerated by the potential and attain sufficient energy to ionise the gas molecules. The current due to the ionized gas molecules adds to the electron current and is proportional to the pressure. To increase the sensitivity a magnetic field is applied, which causes the electrons to travel in long spiral paths. This increases the probability of their colliding with the molecules of rarefied gas. A typical Penning gauge covers the range from 1 to 10^{-6} Pa (10^{-2} to 10^{-8} mbar).

8.1.7 Vacuum components

Standard sizes of copper pipe and associated solder fittings, made for water plumbing, are normally employed to construct the interconnections in basic vacuum systems like that shown in Figure 8.3. For more specialised ultra-high-vacuum applications welded stainless steel tubing is superior. Vacuum components at room temperature, valves, pipe coupling flanges and plastic O-ring seals can be bought. Where flexible joints are required to overcome alignment problems or to provide vibration isolation, metal bellows may be used, but these are expensive. In non-critical applications, where a poorer ultimate vacuum can be tolerated, thick-walled plastic or rubber tubes suffice to make convenient flexible joins. It is important to allow for the effect of the interconnecting pipes on the pumping speed of a vacuum system. There is no point in paying large sums of money for big, fast pumps and then restricting the speed with inadequately sized pipework. However, oversized pipes are cumbersome and the associated valves and couplings more expensive.

The flow of gas in a pipe may be described as viscous or Knudsen, depending on the relationship between the gas molecules' mean free path, $\bar{l}$, and the pipe diameter, $2a$. If $\bar{l} \ll 2a$, then the flow is viscous and if $\bar{l} > 2a$, the flow is Knudsen (also known as molecular flow). In the latter case the rate of flow is determined by collisions between the gas molecules and the pipe walls instead of intermolecular collisions. Assuming an ideal gas of hard spheres, diameter d, the mean free path can be worked out using kinetic theory and is given, in metres, by

$$\bar{l} = \frac{kT}{\sqrt{2}Pd^2},$$

where P is the gas pressure and T the temperature. Typically, d is of order 2×10^{-10} m, and so at room temperature

$$\bar{l} \approx \frac{0.023}{P} m.$$

Therefore in pipework, radius a, it is reasonable to describe the flow as viscous when $Pa \gg 0.01$ and Knudsen when $Pa < 0.01$. In 22-mm vacuum plumbing the flow is Knudsen at pressures less than about 1 Pa. Viscous flow is described by the Poiseuille equations; the mass flow rate, $\dot{M}$, through a pipe, length L, is given by

$$\dot{M} = \left(\frac{m\pi a^4}{16\eta RTL}\right)[P_1^2 - P_2^2],$$

where m is the molecular weight of the gas, η the viscosity of the gas and $P_1 - P_2$ the pressure difference across the ends of the pipe. Assuming the gas to be ideal, the volume flow rate at any point, pressure P, along the pipe is given in terms of $\dot{M}$ by

$$\dot{V} = \left(\frac{RT}{mP}\right)\dot{M}.$$

The low-pressure end of the tube is connected to the pump and the volume flow rate at this point is equal to the speed, S, of the pump, therefore

$$P_2 = \left(\frac{RT}{mS}\right)\dot{M}.$$

Substituting for P_2 in the expression for $\dot{M}$ and rearranging, it follows that the volume pumping speed at the far end of the pipe

$$\dot{V} = S\left(\frac{8\eta LS}{\pi a^4 P_1}\right)\left\{\sqrt{1 + \left(\frac{\pi a^4 P_1}{8\eta LS}\right)^2} - 1\right\}.$$

If the pump is sufficiently fast that $S \gg \pi a^4 P_1/8\eta L$, then the above expression simplifies to

$$\dot{V} \approx \frac{\pi a^4 P_1}{16\eta L}.$$

In this case the speed is limited by the pipe size. The quantity $\pi a^4 P_1/16\eta L$ is known as the conductance of the pipe. If, on the other hand, the pipe is made large so that a $\pi a^4 P_1/8\eta L > S$, then $\dot{V} \approx S$, i.e. the overall pumping speed is limited by the pump.

The mass flow rate in the Knudsen regime is given by

$$\dot{M} = \left(\frac{4a^3}{3L}\right)\sqrt{\left(\frac{2\pi m}{RT}\right)}[P_1 - P_2]$$

(Dushman, 1962). Following the same procedure as for viscous flow, above, the volume flow rate at the far end of the pipe is obtained:

$$\dot{V} = \frac{SC}{S + C},$$

where, in this case, the conductance of the pipe

$$C = \left(\frac{4a^3}{3L}\right)\sqrt{\frac{2\pi RT}{m}}.$$

It follows that, if $S \gg C$, the overall pumping speed is equal to C. Figure 8.4 shows the conductance of various sizes of round cross-section vacuum tube for helium gas, $\eta = 1.3 \times 10^{-4}\,\mathrm{N\,m\,s^{-1}}$ at 300 K.

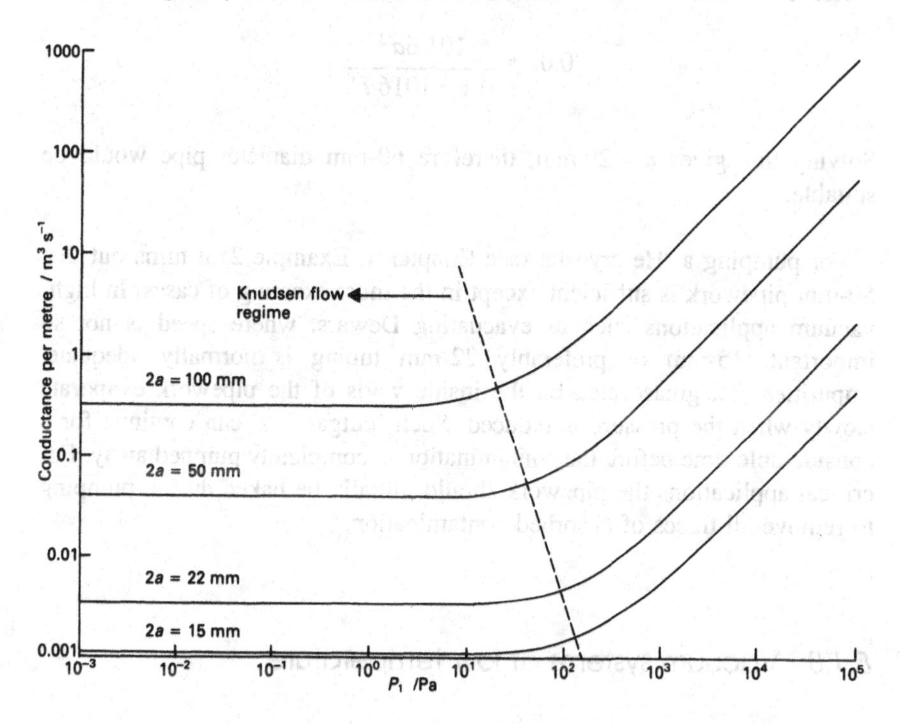

Figure 8.4 *Conductance per metre length of tubes of various diameters as a function of pressure, for helium gas ($\eta \approx 7.5 \times 10^{-6}\,\mathrm{T}^{1/2}\,\mathrm{N\,m\,s^{-1}}$) at T = 300 K*

■ EXAMPLE

A pumping speed of $0.02\,\mathrm{m^3\,s^{-1}}$ at a pressure of 0.25 Pa is required in order to operate a pumped ^{3}He cryostat at a temperature of 0.3 K with a cooling power of 20 μW. A pump having a speed of 6000 l/min is mounted at a 3-m pipe run from the cryostat. What is the minimum diameter of the pipework required to connect the pump to the cryostat?

First, it is necessary to deduce the flow regime relevant to the situation. A glance at Figure 8.4 reveals that at a pressure of 0.25 Pa it would be reasonable to assume that the flow is Knudsen in any sensible sized pipe.

The minimum pumping speed required at the cryostat is

$$\dot{V} = 0.02 = \frac{SC}{S+C},$$

where $C = 1016a^3$. Now $S = 6000\ \text{l/min} = 0.1\ \text{m}^3\ \text{s}^{-1}$, therefore

$$0.02 = \frac{101.6a^3}{0.1 + 1016a^3}.$$

Solving this gives $a = 29$ mm, therefore 60-mm diameter pipe would be suitable.

For pumping a ^{4}He cryostat (see Chapter 4, Example 2) it turns out that 50-mm pipework is sufficient except in the most exacting of cases. In high-vacuum applications such as evacuating Dewars, where speed is not so important, 15-mm or preferably 22-mm tubing is normally adequate. Impurities, i.e. grease etc., on the inside walls of the pipework evaporate slowly when the pressure is reduced. Such 'outgassing' can continue for a considerable time before the contamination is completely pumped away. For critical applications the pipework should initially be baked during pumping to remove all traces of absorbed contamination.

8.1.8 Vacuum systems at low temperature

Ultimately the vacuum system has to reach into the low-temperature regions of the apparatus. Thin-walled, stainless steel, vacuum pipes minimise heat leaks, but must not be so thin that they collapse under vacuum. Because the mass flow rate through a pipe is increased at lower temperatures, the diameter of the pumping line may be stepped down within the cryostat where the temperature is lower. In the viscous flow regime (see Section 8.1.7) the mass flow rate is proportional to $T^{-3/2}$. Therefore, if at room temperature 50-mm diameter pipework is needed then 30-mm and 10-mm pipes would be adequate at 77 K and 4 K respectively. Soldered or welded vacuum joints pose few problems at low temperatures, although the stresses set up upon cooling can fracture a poor or weak joint. Demountable vacuum joints which will operate at low temperature are, however, more difficult to make. Plastic O ring seals are useless at low temperatures, where they contract, freeze and eventually break up. The usual solution is to employ indium metal O-rings. Indium is quite soft at room temperature and when compressed tightly between two surfaces it fills any gaps between them, making a reliable gas-tight seal. Figure 8.5 shows a typical indium seal which retains its integrity to the lowest temperatures. It is made by squashing a thin (approximately 1-mm diameter) indium wire between the flanges. Care should be taken to ensure that the ends of the indium wire overlap and that the bolts are tightened evenly and progressively around the flange to keep the mating faces parallel.

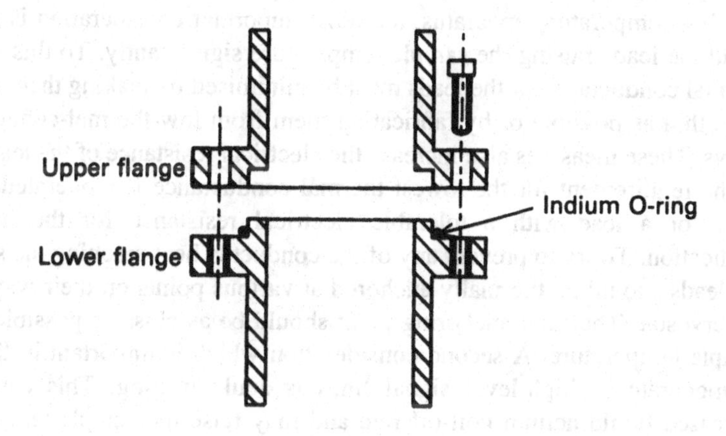

Figure 8.5 *Low-temperature vacuum flange sealed by an indium O-ring. The O-ring is made from indium wire approximately 1 mm in diameter. It is wrapped around the flange and overlapped at the ends to form an unbroken seal. The surfaces and indium wire should be cleaned well before assembly*

Although a poor vacuum joint may appear to be all right when tested at room temperature, it may well be permeable to superfluid helium. Superleaks, as they are known, which become apparent only below the λ-point, can prove difficult to locate using a leak-detector. If one is suspected it is usually necessary to re-make all joints in the offending part of the apparatus.

Often it is necessary to pass electrical leads through a metal flange into a vacuum space, yet maintain their electrical isolation through appropriate insulation. Glass-to-metal feedthrough vacuum seals can be bought for this purpose. A simple, cheap, alternative procedure is to feed a suitable lead (e.g. bare copper wire) through a small hole in the flange and then seal the hole with one of the low-temperature epoxy resins that are readily available.

8.2 Electrical wiring at low temperatures

Many experiments carried out at low temperatures require some form of electrical connection to be made to the sample. Electrical wiring is also needed for resistance thermometers and other devices, e.g. superconducting electromagnets. A wide variety of electrical inputs ranging from direct current to microwave frequency signals and from microvolts to kilovolts may be encountered. High-field superconducting electromagnets may require currents in excess of 100 A.

8.2.1 Direct current and low-frequency wiring

In choosing the leads to carry low-frequency (D.C. to about 30 kHz) inputs into low-temperature apparatus, the most important consideration is how to avoid the leads raising the sample temperature significantly. To this end the thermal conductance of the leads must be minimised by making them as long or as thin as possible or by fabricating them from low-thermal-conductivity alloys. These measures also increase the electrical resistance of the leads, and so the requirement for the lowest thermal conductance is moderated by the need for a lead with a tolerable electrical resistance for the intended application. To try to prevent any of the conducted heat reaching the sample, the leads should be thermally anchored at various points on their way down the cryostat. The final anchoring point should be as close as possible to the sample temperature. A second consideration which is important in the case of moderate to high-level signal lines is Joule heating. This causes an increased liquid-helium boil-off rate and may raise the sample temperature slightly. In some cases, particularly where the leads carry relatively high currents for long periods of time, a compromise between thermal conductivity and Joule heating must be sought (see Section 4.4).

For many small-signal applications, enamelled wires are sufficiently well insulated and the enamel insulation stands up well to low temperatures. Where improved insulation is required, PTFE sleeving is the best choice. It should be a loose fit over the wire to allow for the differential contraction. Insulating materials like PVC become very brittle at low temperatures and should be avoided. Enamelled wires are readily thermally anchored by wrapping them around a cold metal part of the apparatus and sticking them down with a thin layer of insulating varnish. If it is important that electrical leads also form part of the thermal contact to a device, e.g. the leads to a temperature-sensing resistor, then the use of ordinary electrical solder to attach the wires to the device should be avoided. Tin–lead solder turns superconducting and forms a low-thermal-conductivity barrier between the wire and the device. Superconducting transitions in soldered connections may also affect sensitive measurements if they occur somewhere within the range of temperatures over which the measurements are made. Special non-superconducting solders based on a cadmium–zinc alloy are available for low-temperature applications. In making up leads for low temperatures, care should also be taken to minimise as far as possible any stray thermal e.m.f.s that are due either to temperature gradients or to connections between different metals. These thermal e.m.f.s could turn out to be larger than the wanted signal. At the top of the cryostat, electrical leads are normally brought out via a vacuum-sealed multi-way plug.

High-current leads, e.g. superconducting magnet leads, can pose particularly serious problems. Either they must be so thick that the helium boil-off due to thermal conduction is intolerable or the Joule heating is very high. One way of alleviating this problem is to make the lower sections of electrical leads that are immersed in liquid helium from superconducting wire, having low thermal conductivity and zero Joule heating. However, as the helium level falls, any part of the superconducting lead that turns normal may have a relatively high resistance and serious Joule heating can take place. Composite leads, consisting of a thin copper core cladded with a superconductor, are a good compromise. When the cladding is superconducting, it electrically short-circuits the copper core, carries the current with zero Joule heating and contributes relatively little to the thermal conductivity of the lead. When the superconductor is normal it is effectively short-circuited by the copper. Superconducting leads are also useful for sample wiring in highly critical, sub-millikelvin and microkelvin applications, where it is of the utmost importance to have a vanishingly small heat leak. The easiest way to make a composite wire, capable of carrying a few hundred milliamperes, is to tin copper wire with ordinary electrical solder, which is a superconductor at liquid-helium temperatures. Tinned copper wire suitable for this purpose can also be bought.

8.2.2 Radiofrequency lines

High-frequency signal lines that are suitable for low-temperature applications pose greater problems. To prevent interference being picked up by wires carrying small signals and to minimise signal losses by radiation, twisted pairs of wires or coaxial screened cables are essential. Twisted pairs, sometimes called balanced lines because radiation is suppressed if equal and opposite currents flow on each wire, are a practical proposition up to frequencies of a few hundred kHz. At higher frequencies, up to a few GHz, coaxial cables are satisfactory. In this latter case, the electric fields associated with the signal on the inner conductor are completely enclosed by the screen. Furthermore, the outer conductor shields the centre from external, interfering, fields. Coaxial cables are often known as unbalanced transmission lines. An important property of transmission lines is their characteristic impedance, Z_0.

The characteristic impedance of a cylindrical coaxial line is given, in ohms, by

$$Z = \left(\frac{138}{\sqrt{\varepsilon}}\right) \ln\left(\frac{D}{d}\right),$$

where ε is the dielectric constant of the insulating material, D is the inside diameter of the outer conductor and d the outside diameter of the inner conductor. If a coaxial cable is terminated with a load impedance $R_L = Z_0$ then all of the power travelling down the line will be transferred to the load. If $R_L \neq Z_0$ then some of the power will be reflected back up the line. Most modern laboratory equipment is designed to be interconnected via 50-Ω coaxial cables.

Both types of transmission line pose problems of thermal conduction and anchoring when used at low temperatures. Balanced lines work poorly when placed close to other conductors, as is the case when they are wrapped tightly around part of the cryostat for thermal anchorage. The outer conductor of a coaxial cable is easily thermally anchored to metal parts of the cryostat without affecting the performance of the cable. In many cases it is desirable that both the cable screen and the cryostat are at signal ground (0 V). In such circumstances they may be soldered directly together to make an excellent thermal contact. It is the centre conductor that poses the most serious problem concerning thermal anchoring. The insulating material in the cable has a rather poor thermal conductivity compared with that of the metal. Any heat conducted down the centre will therefore be carried directly to the sample.

The normal measures to minimise conducted heat, using the minimum possible thickness of wire, or alloys having a moderately high thermal resistivity, are applicable to coaxial cables. However, these measures increase the electrical resistance of the wire and so degrade the high-frequency performance of the cable. An alternative method of reducing heat leaks takes advantage of the so-called skin effect. High-frequency current is carried only in a thin outer skin of a wire. The skin depth, δ, is given by

$$\delta = \sqrt{\frac{2\rho}{\omega\mu}},$$

where ω is the angular frequency of the signal, μ the magnetic permeability of the wire material and ρ its electrical resistivity. In good conductors, δ is typically of the order of a few micrometres at 100 MHz. On the other hand, heat is conducted uniformly throughout the cross-sectional area of a wire. A significant reduction in the thermal conductance of a cable can therefore be achieved without degrading the high-frequency performance by using a central conductor of tubular cross-section or a core of relatively low-thermal-conductivity material plated with a very thin outer skin of a good conductor like silver.

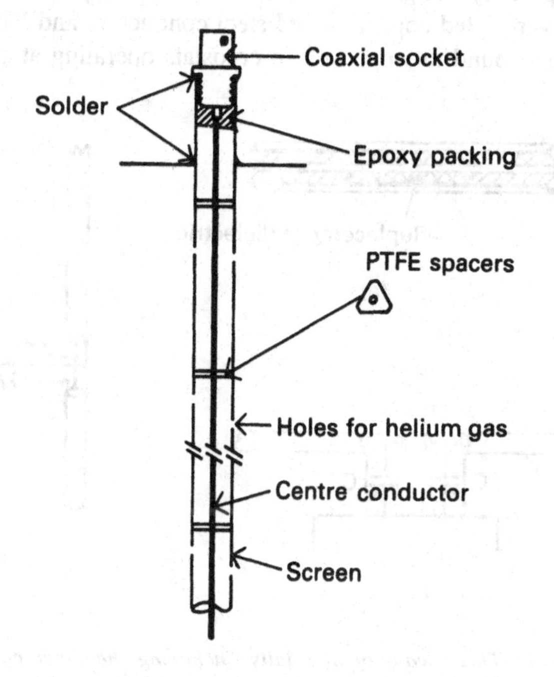

Figure 8.6 *Open coaxial line which is cooled by the flow of helium gas. Standard coaxial connectors are not guaranteed leak-tight, and so the top of the line is packed with epoxy to form a vacuum seal. Alternatively, leak-tight connectors can be purchased, but these work out rather expensive*

Coaxial cables of standard 50-Ω characteristic impedance designed for low-temperature applications can be bought from specialist suppliers. These are rather expensive and are often of semi-rigid construction, making installation into the cryostat somewhat difficult. Sometimes the flow of helium gas exiting the cryostat is exploited to cool the inner conductor of an 'open' coaxial line, as in Figure 8.6. The stainless steel tube forming the centre conductor is spaced apart from the screen by triangular plastic (PTFE) spacers, which allow the passage of cold helium gas through the cable. Open lines are rigid and must be designed into the cryostat from the start. Unfortunately, lines of stainless steel construction are rather lossy at high frequencies. Short sections of coaxial line operating at 4.2 K or below can be fabricated from suitable superconducting wire. The easiest way to do this is to dismantle a length of commercial coaxial cable and replace the inner conductor with a wire of a suitable superconducting material, e.g. NbTi. Where R.F. wiring is to be installed in an existing cryostat it would be desirable to use standard, flexible coaxial cable. Normal copper–polyethylene cables like URM43 have too great a thermal conductivity. Thinner types like URM95 offer some improvement. Sub-miniature coaxial (MCX) cables have been especially designed for *high*-temperature applications. These are made from silver-plated copper coated steel conductors and PTFE insulation, and have been found to work well in cryostats operating at or above 1 K.

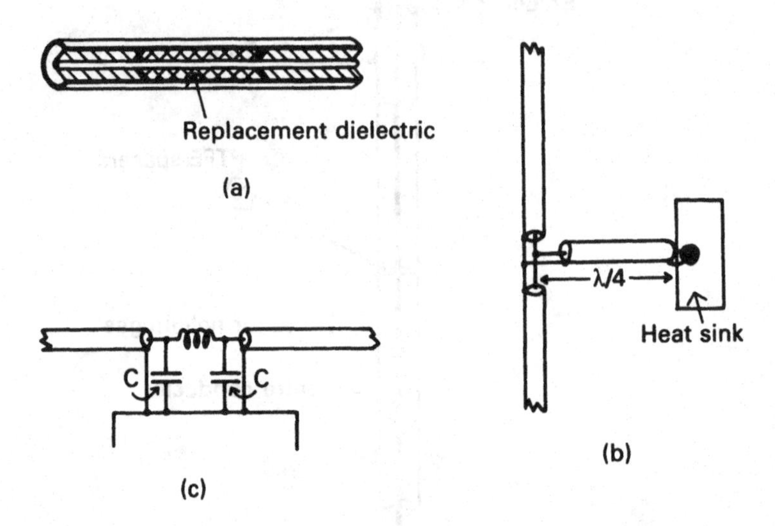

Figure 8.7 *Three ways of thermally anchoring the inner conductor of a coaxial line: (a) replacing a short section of the dielectric with one of higher thermal conductivity, (b) using a short-circuited quarter-wavelength line (only useful when working at a single, very high frequency), and (c) inserting a low-pass π-network filter consisting of silvered-mica capacitors (useful over a band of frequencies up to about 100 MHz)*

For access to lower temperatures, the coaxial cable has to be broken to allow the inner conductor to be anchored, with obvious consequences for its R.F. performance. A number of methods of anchoring the centre conductor are shown in Figure 8.7. One method, shown in (a) involves removing the insulation from along a short length of the inner conductor and replacing it with one of the proprietary epoxy resin materials designed for use at low temperatures. These have a rather higher thermal conductivity than the normal insulating material. If the installation is to operate at one frequency only then the method shown in (b) may be tried. A one-quarter wavelength of coaxial cable is connected in parallel with the line. The end of the λ/4 section is bonded electrically to the cryostat metalwork. This makes a good thermal contact but, because the R.F. impedance at the open end of a short circuited λ/4 length of cable is very high, the signal is unaffected. This method is only effective of course, if the λ/4 section is short compared with the rest of the line, which restricts its use to the higher frequencies. If there is a maximum frequency at which the line is to be used, the π-network filter shown in (c) may be incorporated into the line. The values of the components are chosen according to the following equations:

$$L = \frac{Z_0}{\pi f_c}; \qquad C = \frac{1}{2\pi f_c Z_0},$$

where Z_0 is the characteristic impedance of the coaxial cable and f_c the highest frequency to be carried. The capacitors form the thermal contact and should be of the silvered-mica type. Improved thermal isolation can be achieved by winding the inductor from superconducting wire. Component values yielding a cut-off frequency of 100 MHz when inserted into a 50-Ω coaxial line, are $C = 32$ pF and $L = 0.16$ μH. The inductor can be made by winding 6 turns of wire on a 6-mm drill shank and spacing it to a total length of 6 mm. At 4.2 K this arrangement would be able to sink about 150 μW. Non-superconducting solder should be used for all joints providing thermal anchoring. The π-network also excludes interference from unwanted high-frequency signals which might, if they reach the sample, also result in excess heating.

It is often necessary to make gas-tight seals in coaxial cables where they pass into the cryostat and evacuated sample spaces. In semi rigid cryogenic coaxial line, the dielectric seals the space between the two conductors. An external seal, where the line passes through a hole in a vacuum flange, can be made by soldering or packing out the hole with epoxy. A low-melting point solder must be used to avoid heat damage to the dielectric. The ends of commercial coaxial cables can be sealed by removing the standard insulating materials from a short length and replacing them with a bead of epoxy or silicone rubber compound.

8.2.3 Waveguides at low temperatures

At much higher microwave frequencies, coaxial cables become too lossy and so waveguides are used to carry the signals. Waveguides commonly take the form of a hollow tubular conductor of either circular or rectangular cross-section, which confines the microwave fields to propagate in one dimension. The cross-sectional dimensions of the tube are related to the narrow band of microwave frequencies it is intended to carry. Stainless steel is too lossy to use as a cryogenic waveguide in critical applications. However, plating the inside surface with a thin layer of a good conductor such as silver improves the performance. Commercial copper, brass or nickel silver waveguide sections are satisfactory if care is taken to thermally anchor the guide at many points in the cryostat. This is done in the same way as for pumping and support tubes, by soldering the guide into holes in baffles etc. The waveguide can provide part of the mechanical support for the internal cryostat structure, and with careful design can also be used as a pumping line. The corresponding reduction in the number of other tubes helps to compensate for the increased helium boil-off due to the waveguide. Another source of unwanted heat input associated with waveguides is radiation. The smooth conductive walls of the guide are highly reflective to infrared radiation entering at the high-temperature end. Bends at the low-temperature end of the guide help to reduce radiation reaching the sample, but there is not always space in a cryostat to do this. An alternative method of reducing infrared radiation is to insert low-density insulating plugs into the guide at various points. A waveguide can be vacuum-sealed by inserting a thin plastic washer between coupling flanges at room temperature. The mating surfaces should be smeared with a film of low-vapour-pressure vacuum grease before assembly. Waveguide flanges can be made vacuum-tight at low temperatures by assembling them with a thin indium O-ring seal between them.

Often the waveguide ends at a low-temperature microwave cavity in which the sample is mounted. The tuning of a cavity is critically dependent on its dimensions, which change as it is cooled down. The normal procedure is to design the cavity to be on-tune at room temperature and include a means of fine tuning the cavity when it is cooled. The tuning apparatus is actuated from room temperature via a low-thermal-conductivity mechanical link.

8.3 Optical access

Light access to a sample in a cryostat, at low temperature, may be provided by means of a light pipe or optical fibre running from the top of the cryostat or by windows in the tail of the cryostat. One problem associated with both methods of optical access is that, in addition to the wavelengths of interest, unwanted radiation may reach the sample and increase the heat loading. To alleviate this problem, light pipes or windows should be made as small as possible consistent with adequate optical access. If the experiment is to be performed using only a narrow range of wavelengths then it is helpful to use windows or fibres having an appropriately narrow passband. A number of window materials for covering different bands, i.e. ultraviolet, visible and infrared, are commercially available.

Quartz is a good general-purpose cryogenic window material in the range 1 K to room temperature. It is transparent to ultraviolet, visible and near-infrared wavelengths and chemically stable. The main disadvantage with quartz is that it has a few strong absorption bands in the near infrared.

Calcium fluoride has a smooth optical passband, ranging from the ultraviolet to the far infrared, and is non-hygroscopic at room temperature. The disadvantages with this material are that it is easily scratched and, because it has a high coefficient of thermal expansion and low thermal conductivity, it is highly susceptible to thermal shock if the temperature changes rapidly. For windows with a passband extending from the visible red into the extreme infrared (up to about 35 μm), a synthetic mixed crystal of thallous bromide and iodine, KRS5, is a good choice. It has a fairly constant transmission factor of better than 60 per cent over its passband. Other window materials suitable for the 1–300 K temperature range include sapphire and clear polyethylene terephthalate (Mylar) film. Thin sheets of Mylar also make good X- and γ-ray windows. They may be coated with a thin aluminium layer to exclude unwanted visible and infrared wavelengths which may place an additional heat loading on the sample. Mylar is slightly porous to helium gas, which may cause problems if it is used as a window on a helium cell. For work involving only a very narrow range of optical wavelengths, windows can be coated with materials which restrict their passband, and so prevent unwanted radiation reaching the sample.

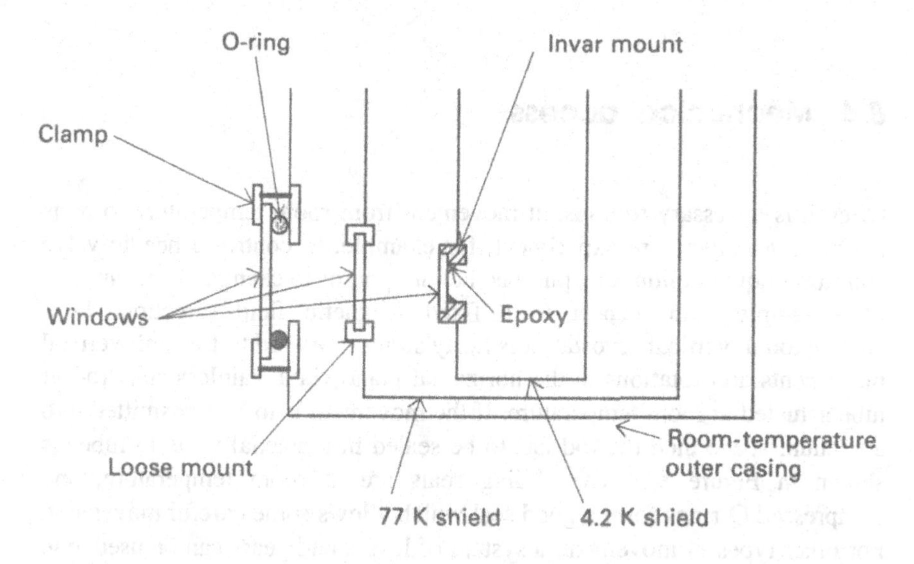

Figure 8.8 *Section of the tail of an optical cryostat showing the methods used to mount windows. The 77-K radiation shield does not hold a vacuum. Quartz is a good general-purpose window material*

Windows must be positioned wherever light has to pass through the walls of a vacuum can on its way to the sample (Figure 8.8). They may also be placed in radiation shields, but a simple hole will usually suffice. Room-temperature windows are made-vacuum tight simply by clamping them between plastic O-rings. More care is needed when mounting windows at low temperatures. Upon cooling, many window materials contract rather less than metals. The stresses set up between a window and its mount may be sufficient to fracture the window or vacuum seal or, at the very least, affect the optical properties of the window. Invar has a low coefficient of thermal expansion and is suitable for making low-temperature window mountings. The mount should be strong enough to withstand the forces resulting from thermal contraction of the surrounding metalwork without distorting. A vacuum seal may be made by gluing the window into the mount or by gently clamping it between indium O-rings. In designing an optical cryostat, allowance must be made for the thermal contraction of the inside of the cryostat, which will cause a vertical misalignment of the inner and outer windows. Because it is not a straightforward matter to design and build a reliable optical access cryostat, many experimenters choose to buy one.

Low-cost, plastic optical fibre is suitable for illuminating samples at low temperature. To allow for thermal contraction, sufficient slack must be left in the fibre at room temperature. The main problem with fibres is their rather narrow optical passband.

8.4 Mechanical access

Often it is necessary to transmit movement from room temperature to parts of the low-temperature experiment, for example, to control a needle valve admitting liquid helium to a pumped helium pot, or to change the orientation of a sample with respect to a fixed magnetic field direction. In a conventional, vertical cryostat, it is fairly straightforward to transmit vertical movements and rotations in the horizontal plane via a stainless steel rod or tube actuated at room temperature. If the movement is to be transmitted into a vacuum space then the rod has to be sealed in a coaxial vacuum tube, as shown in Figure 8.9. The sliding seals are at room temperature, and compressed O-rings form a good seal which allows some careful movement. For other types of movement, a system of levers and gears can be used, e.g. a worm gear may be used to change the plane of rotation. There are two problems to watch out for: conventional lubricants cannot be used because they freeze solid, and thermal contraction can introduce sloppiness into the system. It is not always possible to see the sample, and so some sort of position indicator is required. In some cases this has to be mounted close to the sample to avoid errors due to twisting in the long stainless steel rod and the position information relayed electrically. Experiments involving sample motion are only really practical at temperatures down to about 1 K. At much lower temperatures, frictional heating effects become intolerable.

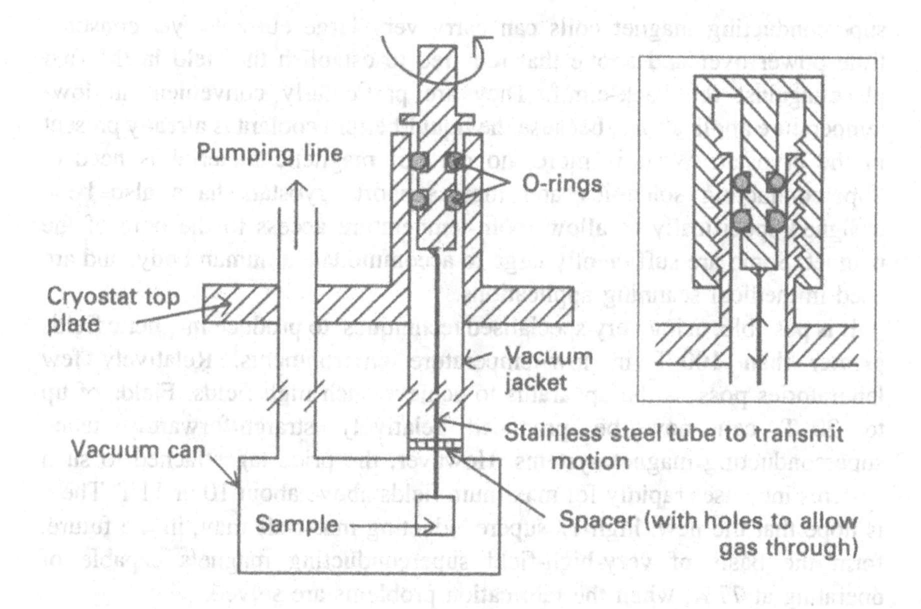

Figure 8.9 *Simple arrangements for transmitting horizontal-plane rotational, and (inset) vertical movements to a sample within a vacuum can at low temperatures. Most metals retain enough flexibility at low temperatures for wires etc. to be run to the sample, provided enough excess length is allowed to cater for the movement*

8.5 High magnetic fields and low temperatures

Magnetic fields of up to about 2 T have long been produced by large, iron-cored electromagnets operating at room temperature. To conduct experiments at low temperatures in fields produced by such magnets, the tail of the cryostat has to be made narrow enough to fit in the gap between the poles. This restricts the working volume and complicates the cryostat design. The size of the magnet goes up very rapidly with working volume, and because of the resistance of the windings such a magnet may consume enormous electrical power and require copious quantities of cooling water. Another aspect is that it may be required to direct light along the magnetic field, direction which is awkward, requiring holes in the iron cores or mirrors in the gap. Fortunately, high magnetic fields and low temperatures go hand in hand through superconductivity. 'High' fields, in this context, are those of the order 1 T and greater. Such fields are used for magnetic cooling and in experiments on the fundamental properties of matter at low temperatures, e.g. the quantum Hall effect in semiconductors and the magnetic susceptibility of dilute ^{3}He–^{4}He solutions (see Sections 2.2.4 and 3.3). The windings of superconducting magnet coils can carry very large currents, yet consume little power over and above that required to establish the field in the first place against the back-e.m.f. They are particularly convenient in low-temperature applications, because the liquid-helium coolant is already present in the cryostat. What is more, no core of magnetic material is needed. Superconducting solenoids and their support cryostats have also been designed specifically to allow room-temperature access to the bore of the magnet. Some are sufficiently large to accommodate a human body, and are used in medical scanning applications.

It is possible, using very specialised techniques, to produce magnetic fields greater than 100 T in low-temperature environments. Relatively few laboratories possess the apparatus to achieve such high fields. Fields of up to 20 T can now be produced relatively straightforwardly using superconducting magnet systems. However, the price tag attached to such systems increases rapidly for maximum fields above about 10 or 11 T. There is hope that the new, high-T_C superconducting materials may, in the future, form the basis of very-high-field superconducting magnets capable of operating at 77 K, when the fabrication problems are solved.

Superconducting electromagnets are normally wound from type-2 superconducting wire, i.e. NbTi, which maintains its zero-resistance state in the presence of a strong magnetic field. Certain other alloys of niobium, such as Nb_3Sn, have a higher value of upper critical field (above which there is a transition back to normal conductivity), but are less ductile than NbTi and so are harder to fabricate into magnets. A single filament or many small filaments of the superconductor are embedded in a normal metal, usually copper, which provides good thermal stabilisation and mechanical strength. For a particular value of overall diameter and magnetic field, multi-filament wires have a higher value of critical current than a single-filament wire. The wire is coated with insulating enamel and wound under moderate tension onto a former. The former needs to be rigid to withstand the high mechanical forces that can be generated during operation of the magnet. Formers are often machined from a solid block of aluminium or plastic. After winding, the coils are sometimes potted in epoxy for extra rigidity. The most common magnet geometry is the solenoid, which is the easiest to fabricate and, if properly designed, the most efficient, in terms of the ratio of maximum possible field strength to stored energy. The disadvantage of the solenoid geometry is that it only allows access to the maximum field at its centre from the ends. An alternative magnet geometry, which allows access to its maximum field region from three mutually perpendicular directions, is the so called split pair. It is particularly useful in some optical experiments, where light access in a direction perpendicular to the field is required. The typical split-pair electromagnet consists of two identical coils on the same axis, separated by a few centimetres and wired in series. The maximum field is produced on the axis, mid-way between the coils. Split-pair magnets are considerably more bulky than solenoids for the same value of maximum achievable field strength.

The magnetic flux density in the centre of a solenoid, length l and average diameter D, is given approximately by

$$B \sim \frac{\mu_0 N I}{(l^2 + D^2)^{1/2}},$$

Where N is the number of turns and I the current (see, e.g., Bleaney and Bleaney, 1976). This expression does not take into account fringing fields and the finite thickness of the windings. The design of modern magnets is carried out using computers to solve the relevant fundamental electromagnetic equations and so to model the magnet properties. A typical, small, superconducting solenoid 20 cm long, of 10 cm outside diameter and 2.5 cm bore, consists of about 60 000 turns and produces a maximum field of about 8 T at a current of about 80 A. The self-inductance of the coil is quite high, about 14 H for the example given above. A simple calculation shows that the energy stored in the magnet at maximum field is 44.4 kJ, sufficient to run a one-bar electric fire for almost a minute!

The magnet power supply needs to be capable of providing a stabilised current of up to 100 A or more, depending on the requirements of the magnet. The voltage at its output terminals must be sufficient to overcome resistive voltage drops in the magnet leads and the back-e.m.f. across the magnet when the field is being increased. These typically amount to a couple of volts in total. The power supply must also be able to absorb the energy released when the magnetic field is reduced. The power supply will normally incorporate some automatic means of sweeping the current up and down gradually so that the back-e.m.f., which is proportional to the rate of change of current, does not get too large. The magnet leads are made of heavy-gauge wire to carry the high current without significant voltage drop or heating. Such leads inevitably conduct significant heat, and so every effort is taken to ensure that the leads are effectively cooled by the helium gas inside the cryostat. To help cooling, the leads have a large surface-area-to-volume ratio, i.e. they are made from flat tape or hollow tubes. If the magnet is operated in persistent mode (see below) then the magnet leads are energised for relatively short periods, and so may be fabricated from materials having a lower thermal conductivity than copper. Alloys such as brass might be used, at the expense of increased helium loss owing to Joule heating during the periods while the leads are energised. Sections of the lead immersed in liquid helium may be fabricated from superconducting wire or a composite normal–superconducting cable.

Field stability and liquid-helium consumption are improved by operating the magnet in persistent mode. A persistent-mode magnet has a superconducting short-circuiting link across the coil terminals. The link is in thermal contact with an electric heater. When the heater is on, the link is normal and the supplied current flows preferentially through the zero-resistance magnet windings. Once the required field is established, the heater is switched off and the link becomes superconducting. In effect, the magnet and link now become a closed superconducting loop with a persistent, circulating current trapped in it. The supply current may be totally removed from the magnet leads and yet the field remains constant.

Because of the large amounts of energy stored in the magnetic field, an energised magnet must be treated with care. If part of the magnet goes normal then the heat generated in it will drive neighbouring regions normal. More heat is generated in these, and so on, until most of the energy stored in the magnetic field is dissipated. A sudden collapse of the field in this way is called a quench. It is accompanied with a sudden rush of evaporated helium and the possibility of high voltages being induced in the magnet windings. Quenches are caused by:

1 Increasing the current beyond the value corresponding to the maximum rated field of the magnet.

2 Allowing the liquid-helium level to drop below the top of the magnet.

3 Increasing the field too rapidly.

4 Disconnecting the power supply while the magnet is energised, if it is not in the persistent mode.

5 Turning on the superconducting switch heater of an energised persistent-mode magnet without the appropriate current being present on the supply leads.

Sudden, small movements of windings or rapid changes in field will induce voltages in the wire. Components of the induced voltage parallel to the current flow cause energy to be dissipated by the wire. If the heat is able to escape without the wire increasing in temperature to above its superconducting transition then all is well; otherwise, a quench is likely to occur. This is why it is important that the magnet is of rigid construction and stable against rapid thermal fluctuations. Should the worst happen then the magnet can be protected from permanent damage, due to arcing between the windings, by connecting low-value, high-power resistors or power diodes across its terminals to dissipate some of the released energy. These are preferably mounted within the cryostat, on one of the baffle plates above the liquid helium. The magnet power supply should *not* be trusted to protect the magnet, because the current leads might break. A pressure release device in the helium return line, near the Dewar exhaust, prevents dangerously high pressures being built up inside the cryostat as a result of the rapid evaporation of the liquid helium. Quench protection for the magnet power supply is normally provided by high-power rectifier diodes connected internally across the output terminals.

It is rarely that a magnet will achieve its designed maximum field the first time it is used. After manufacture, transport or a long period of storage, superconducting magnets need to be 'trained' by repeatedly ramping up the current until they quench. Successive quenches are found to occur at a higher critical field. It is believed that this process causes the windings to mechanically 'settle down'. The point is ultimately reached where further quenching leads to little improvement in the critical field.

It is possible to build small magnets oneself that are capable of generating fields of up to a couple of tesla. The superconducting wire can be bought from cryogenic accessory suppliers. Designing magnets to produce a high field and/or provide highly homogeneous fields is a difficult and costly process. Such magnets are best bought. The field homogeneity is normally specified as a percentage variation of the field over a particular volume at the centre of the magnet. High-homogeneity magnets sometimes have extra 'shimming' coils fitted to correct for variations in the field of the main solenoid. Such magnets are more costly and require a more complicated power supply, but are essential for certain applications, e.g. for nuclear magnetic resonance studies of large-volume samples.

Ferrous materials are avoided in cryomagnetic systems, because their presence causes distortion of the magnetic field. Ferrous parts would also be subjected to strong magnetic forces. Another important consideration, particularly in systems working at millikelvin temperatures and below, is the possibility of unwanted heating due to eddy currents. Eddy currents are induced in conductors that are subject to a changing magnetic field and cause Joule heating of the conductor. The size of the induced current is proportional to the cross-sectional area of the conductor and its orientation relative to the field direction, it being a maximum when the plane of the conductor is perpendicular to the field. Where it is necessary to have large volumes of conducting material inside the magnet, for example in nuclear cooling applications, bundles of insulated thin wires or plates are oriented parallel to the field direction. Conductors in the form of closed rings within the magnetic field have slots cut in the circumference to break the eddy current circuit.

This chapter has only broached the subject of low-temperature experimental techniques. Low-temperature experiments are also carried out on samples under conditions of high hydrostatic pressure and high uniaxial stress, in a rotating reference frame and involving the detection various sub-atomic particles, to mention just a few. The reader who is seriously interested in setting up a low-temperature experiment is referred to books in the bibliography at the end of this chapter. Some of the books also contain details of manufacturers and addresses of suppliers of low-temperature equipment and accessories.

Bibliography

Betts, D. S. (1976). *Refrigeration and Thermometry Below 1 K* (Brighton: Sussex University Press)

Hoare, F. E., Jackson, L. C. and Kurti, N. (1961). *Experimental Cryophysics* (London: Butterworths)

Lounasmaa, O. V. (1974). *Experimental Principles and Methods Below 1 K* (New York: Academic Press)

Richardson, R. C. and Smith, E. N. (1988). *Experimental Techniques in Condensed Matter Physics at Low temperatures* (California: Addison Wesley)

Rose-Innes, A. C. (1973). *Low Temperature Laboratory Techniques* (London: English Universities Press)

White, G. K. (1959). *Experimental Techniques in Low-Temperature Physics* (Oxford: Clarendon)

On vacuum technique

Dushman, S. (1962). *Scientific Foundations of Vacuum Technique* (New York: John Wiley)

Appendix: Laser cooling

It might seem surprising that gaseous atoms may be cooled by irradiating them with a powerful laser. However, such optical cooling has produced some of the lowest kinetic temperatures yet attained (about 2.5 μK using caesium atoms).

The relationship between the kinetic velocity of gas atoms and absolute temperature was considered in connection with the ideal gas thermometer described in Chapter 1. The mean kinetic energy of the atoms is proportional to the absolute temperature. In fact, for a monatomic gas in three dimensions, the equipartition theorem gives

$$\overline{KE} = \frac{3kT}{2},$$

The mean kinetic energy is equal to $m\overline{v}^2/2$, where m is the atomic mass and $\overline{v}^2$ the mean square speed. Therefore,

$$T = \frac{m\overline{v}^2}{3k},$$

and so, if it were somehow possible to slow down the atoms of a gas, this would be equivalent to lowering its temperature.

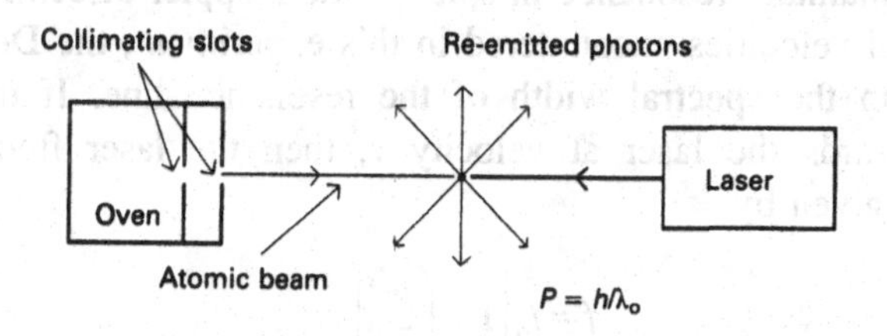

Figure A.1 *Geometry of 1-D optical cooling experiment; photons having momentum in the opposite direction to the atomic beam are absorbed. The re-emitted photons go in all directions, at random*

In the method of laser cooling illustrated in Figure A.1, which was one of the first used, a collimated beam of atoms, emanating from an oven and travelling in a high vacuum, is slowed by the radiation pressure of a counterpropagating laser beam. The wavelength of the laser radiation was tuned so that the photons are resonantly scattered by the atoms. When an atom of mass m and velocity v absorbs a photon from the counterpropagating beam, its forward momentum, mv, is decreased by an amount equal to the photon momentum, which is h/λ, where λ is the laser wavelength. Absorbing a photon raises the atom to an excited energy state. After a short time, the atom spontaneously returns to the ground state, emitting a photon. There is no preferred direction for the emitted photon but the atom recoils in the opposite direction. After a sufficiently large number, N, of collisions, the net result is that the atom's original motion is dissipated and it is left moving in a random direction in three dimensions with an average speed $\sqrt{N}\,v_r$, where v_r is the single-photon recoil velocity. The whole process is rather like trying to stop a cannonball by firing ping-pong balls at it.

As an example, consider the slowing of sodium atoms ($m = 3.82 \times 10^{-26}$ kg), initially at 1700 K in the oven, by resonant scattering of 589-nm (Na D-line) radiation. The average initial momentum of the sodium atoms is $mv \sim (3mkT)^{1/2} = 5.19 \times 10^{-23}$ kg m s^{-1}, while the photon momentum is $p = h/\lambda = 1.13 \times 10^{-27}$ kg m s^{-1}. Therefore, after about 46 000 scatterings an atom will have been 'stopped', except for a small residual velocity, $215v_r = 215p/m = 6.4$ m s^{-1}. The temperature of sodium gas corresponding to this average atomic speed is only 37 mK! If the photon flux is sufficiently high for the scattering process to be saturated, i.e. the time between atomic photon collisions is much less than the characteristic time for the absorption–emission process, then an atom may be stopped in about 1.5 ms after travelling a distance of about 1 m. To be sure that the atoms cannot diffuse out of the laser beam, it is clear that it must have a minimum diameter of about $0.0015v_r = 0.010$ m.

In practice, the main problem encountered when cooling atoms by this method is to maintain resonance in spite of the Doppler effect. At even the lowest thermal velocities encountered in this experiment, the Doppler shift is greater than the spectral width of the resonance line. If the atom is travelling towards the laser at velocity v, then the laser frequency, for resonance, is given by

$$f = f_0\left[1 - \left(\frac{v}{c}\right)\right],$$

where f_0 is the resonance frequency of the atom at rest. As the atom slows, the laser frequency must be swept rapidly to remain on resonance (Ertmer *et al.*, 1985). An alternative method of maintaining resonance as the atom slows is to tune the atomic energy levels, and hence f_0, by varying an applied magnetic field (Phillips *et al.*, 1982).

The Doppler effect can be used to advantage in three-dimensional laser cooling. The already slow atoms are subjected to standing-wave radiation fields along three mutually orthogonal directions, as shown in figure A.2. The frequency of the laser from which the six beams are derived is slightly less than the atomic resonance frequency. There is no scattering force on a stationary atom, owing to the symmetry of the arrangement of beams. However, if an atom has a component of velocity directed against one of the beams, then it blue shifts into resonance with that beam and red shifts out of resonance with the co-propagating beam. The atom, therefore, absorbs more photons from the counter-propagating beam and so experiences an imbalance in the scattering force, which tends to slow its motion. The resulting random-walk velocity in the other directions is slowed by the orthogonal beams, and the atom is 'trapped' in the region where the three beams intersect. The average time taken for a cold atom to diffuse out of the cooling zone is much longer than the time between arrivals from the incoming beam, and so the density of cold atoms is considerably higher than in a 1-D cooled atomic beam. The name 'optical molasses' was coined to describe the collection of very-slow-moving atoms (Chu *et al.*, 1985).

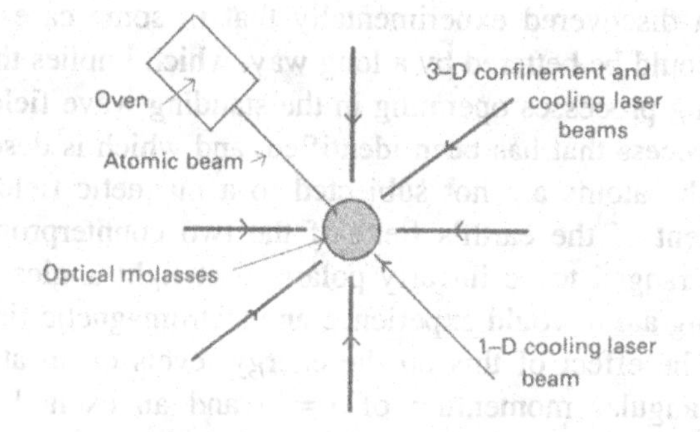

Figure A.2 *Experimental arrangement for 3-D confinement and Doppler cooling of atoms*

Theoretically, 3-D laser cooling is unable to attain the goal of a temperature corresponding to the single-photon recoil energy (2.5 μK in the case of Na). This arises from the finite lifetime, τ, of the excited state of the atom before spontaneous photon emission takes place. Through the uncertainty principle, the atomic resonance has an inherent width, in energy terms

$$\Delta E \approx \frac{h}{\tau} = h\Gamma,$$

where Γ is the spontaneous decay rate. It seems intuitively reasonable, but not at all easy to prove, that the atomic kinetic energy cannot be reduced much below ΔE. The so-called Doppler cooling limit is obtained by expressing this minimum energy in terms of temperature,

$$T \sim \frac{\Delta E}{k} = \frac{h\Gamma}{k}.$$

In the case of sodium, Γ is 10 MHz and the theoretical minimum temperature a mere 240 μK. The fact that such low temperatures can be reached has been confirmed experimentally.

Note that, even if an atom could be found that had an infinitely sharp resonance, 3-D laser cooling would not contravene the third law of thermodynamics. This is because the lifetime of the excited state would be infinitely long, and so it would take an infinite amount of time for the cooling process to work.

It was soon discovered experimentally that in some cases the Doppler cooling limit could be bettered by a long way, which implies that there must be other cooling processes operating in the standing wave field of the three beams. One process that has been identified, and which is described below, requires that the atoms are not subjected to a magnetic field greater than about 2 per cent of the earth's field. If the two counterpropagating laser beams were arranged to be linearly polarised at right angles to each other then the moving atom would experience an electromagnetic field of varying polarisation. The effect of this on the energy levels of an atom, having a ground state angular momentum of J = ½ and an excited state angular momentum of J = 1½, is to periodically lift the degeneracy of ground state levels and modulate the ground state energy. The effect is only small, much less than the Zeeman splitting of the ground state produced by the earth's magnetic field, hence the need for good magnetic screening. Initially the atom is in one of the ground state magnetic levels. As it moves in the radiation field, its potential energy oscillates and so, to conserve energy, its kinetic energy must oscillate in the opposite sense. The atom therefore slows down and speeds up, but, in the absence of resonant photon scattering, this does not result in cooling, because its average energy remains unchanged. Now, if the laser frequency is chosen correctly, the probability of photon absorption is greatest when the ground state energy is near a maximum, i.e. when the atom is moving more slowly. Photon emission, however, favours an atomic transition to a minimum in the ground state energy associated with the lower of the two magnetic levels. After the transition, the atom is moving at the same velocity as it was when the photon was absorbed, the difference in the potential energy being converted to optical radiation instead of kinetic energy. Using caesium atoms, kinetic temperatures down to about 2.5 μK have been achieved, which is just a few times the single-photon recoil limit for these atoms (Salomon *et al.*, 1990). Other, more complicated physical mechanisms have been proposed, which predict that temperatures lower than those corresponding to the single-photon recoil energy could be attained. Using metastable helium atoms, temperatures down to 2 μK (just half the single-photon recoil limit; see Table A.1) have been attained.

Table A.1 Summary of the properties of some of the atoms used in laser cooling experiments

Atom	Mass/kg	Resonance λ/nm	Single-photon recoil limit/μK	Doppler limit/μK	Lowest temperature/ μK
Na	3.82×10^{-26}	589	2.45	240	240
Cs	2.21×10^{-25}	852	0.2	125	2.5
He^*	6.64×10^{-27}	1083	4.1	23	2

Of course, even the highest temperatures produced by these methods cannot be measured conveniently. The gas sample is too small to cool a conventional thermometer by conduction. In practice, temperature is assessed by switching off the confinement laser beams. The cloud of slow-moving atoms then drops under gravity, and as it does so it expands owing to the spread of atomic velocities within it. After a short drop, the cloud passes through a probe laser beam and its fluorescence can be observed. The increase in the size of the cloud is estimated from the spatial extent of the fluorescence signal, and so the r.m.s. velocity of the atoms is determined.

The reader may be wondering why, if even a small thermometer cannot be cooled by this method, the technique is of any use whatsoever. There is the aesthetic attraction of producing a gas in which the atoms move so slowly that quantum-mechanical effects become significant. The thermal de Broglie wavelength of a helium atom at 2 μK is 1.4 μm, which is greater than the wavelength of the confining laser beam, and so it is no longer possible to treat the position of the atom as a classical quantity. It is also possible that the ultracold atoms could undergo Bose condensation, as in the case of liquid helium-II. Practical applications of this technique include high precision-spectroscopy and metrology. At such low atomic speeds, Doppler broadening of the spectral line is negligible, and linewidths approaching the homogeneous limit can be obtained.

References

Chu, S., Hollberg, L., Bjorkholm, J. E., Cable, A. and Ashkin, A. (1985). *Phys. Rev. Lett.*, **55**, 48
Ertmer, W., Blatt, R., Hall, J. L. and Zhu, M. (1985). *Phys. Rev. Lett.*, **54**, 996
Phillips, W. D. and Metcalf, H. (1982). *Phys. Rev. Lett.*, **48**, 596
Salomon, C. *et al.* (1990). *Europhys. Lett.*, **12**, 689

Index

Printed in the United States
By Bookmasters

Printed in the United States
By Bookmasters